from problem
to program

THE
SCHOOL
MATHEMATICS
PROJECT

**computing in
mathematics**

from problem
to program

M. E. WARDLE

CAMBRIDGE
AT THE UNIVERSITY PRESS 1972

Published by the Syndics of the Cambridge University Press
Bentley House, 200 Euston Road, London NW1 2DB
American Branch: 32 East 57th Street, New York, N.Y.10022

ISBNs:

0 521 08301 X hard covers
0 521 09684 7 paperback

Printed in Great Britain by
William Clowes & Sons Limited
London, Colchester and Beccles

THE SCHOOL MATHEMATICS PROJECT

The Project was originally founded in 1961 by a group of practising school-teachers who believed that there were, at that time, serious shortcomings in traditional school mathematical syllabuses, and that there was a need for experiment aimed at developing new syllabuses which would reflect modern ideas and applications of mathematics.

The activities of the S.M.P. are, by now, well-known, as are its series of texbooks — Books A-H for the main school, Books 1-5 and the Additional Mathematics Books for O-level, Advanced Books 1-4 and Further Mathematics Books for A-level. Full details of these publications and activities, and of teacher-training courses, are given annually in the current Director's Report which may be obtained on request to the S.M.P. Office, Westfield College, Kidderpore Avenue, London, NW3 7ST.

So much is, in a sense, past history. But the S.M.P. continues to look forward and to encourage a process of more or less continual change and updating of school mathematics teaching. Thus, among other things, it now turns its attention to the implications that computing power in the school classroom holds for mathematics.

Not enough is yet known about how computers can best be used in the classroom to illumine mathematical curricula: there is much exploratory work first to be done. The books of this new series — under the generic title *Computing in Mathematics* — therefore investigate various applications of computers to school mathematics, and it is hoped that, by provoking discussion over the next few years, they will help to prepare the ground for a more cohesive course in the future.

Finally it is particularly hoped that, as has been the case with previous S.M.P. books, teachers from all types of schools will send their comments and criticisms to the editors of these books.

Westfield College
January 1971

BRYAN THWAITES
Director

This book is based on the original contribution of

M. E. Wardle

and has been edited by B. H. Blakeley.

Grateful acknowledgement is made of the help of those teachers who have submitted comments and criticisms of the manuscript.

The Project owes a great deal to its secretaries, Miss J. Sinfield and Miss J. Try, for their assistance and typing in connection with this book.

Thanks are due to International Computers Ltd for permission to reproduce the photographs on pages 1, 3 and 5.

contents

preface

This book has been written from the author's experience of teaching children aged eleven to eighteen and running introductory courses on programming a computer, for teachers and student teachers.

The book is intended for the beginner and lays the foundation for programming at any level. It shows by means of many worked examples how a problem can be broken down into easy stages and how the list of instructions for implementing them can be written. At the same time it attempts, with the aid of a simple model, to show how the computer reacts to these instructions.

The first chapter outlines the five main components of a computer and indicates the role each plays. The next four chapters introduce and develop the type of instructions that a computer can understand and act on. They cover in detail the basic programming ideas involved in the input, calculation, and output stages and incorporate the use of conditional and unconditional jumps. In addition the ideas of a counter, sum, and raising to a power are discussed.

Chapter 6 explains how subroutines for expressions such as sine x, can be built up and used within a program. The final chapter takes a close look at the various stages in a number of different problems, and considers how the allowable instructions can be combined to enable the computer to solve the problem.

The matchbox computer is developed alongside each new idea and this can be used as an optional project simulating every aspect of the computer, from simple arithmetic to the full stored program facility. Much of the early development of the matchbox computer was originated by J. D. Tinsley.

Experience has shown that most children, up to the age of sixteen, can cope with the work covered in the first four chapters, but that the conditional jump instruction dealt with in chapter 5 should only be introduced when the earlier ideas have been fully grasped. This may produce a natural break as the thinking involved here can often be a major hurdle, which, in many cases is a good indication as to whether an individual can go any further in programming.

Chapter 7 has been included to try and show the type of thinking which a programmer may use and the variety of ways in which a problem can be tackled. It is suggested that this chapter is dipped into, when a problem of the particular type occurs, rather than worked through systematically.

Whilst the main text is not written in a particular commercial language, an appendix has been included to show how the language used can be translated directly into whatever language the reader may have available.

The main objective of the book is to introduce the basic ideas behind programming at a level with which most children can cope. It has deliberately stopped short of many of the subtleties of a high-level language, as too often these are met without any real understanding.

1 the computer

Many people tend to think that a computer is merely a machine which is capable of doing a large number of arithmetic calculations in a very short time. To a certain extent this is true, but at the same time one of its main features is its ability to STORE vast quantities of information which can be made quickly and readily accessible to the user. In addition, unlike a simple calculating device, the computer is able to store the set of instructions which are necessary in order to manipulate and process this information. It is also able to carry out these instructions automatically, with only a minimum amount of assistance from the operator.

figure 1   A computer installation

Let us now consider a fairly simple problem and see just what facilities are required by the computer to give us, in a meaningful form, the information and results that we need.

At the end of each term most form teachers have to collect the marks obtained by each of their pupils in a number of different subjects.

Collect this information and produce a final form order for the particular class.

Even in this situation there is a fair amount of information (or DATA). There are the names of the pupils in the class, the different subjects taken, and also the marks obtained by each pupil in each subject. We also need to know how the various marks are to be combined, whether some subjects such as English and mathematics are more important than others, and what part optional subjects are to play in the final total.

Having obtained this information we will have to calculate the total mark for each pupil, and possibly an average mark. These marks will then have to be sorted from highest to lowest to produce the final form order. Two lists may be needed, an alphabetical one showing each child's mark and place, and also one showing the order of the children with their range of marks.

If, as in many schools, one man, or perhaps the office, is responsible for doing this for every form, you can see the tremendous amount of work that could be involved. A lot of this work would be of a repetitive nature and this is where the computer comes into its own.

Like many problems the one just outlined has three distinct stages:
(1) The original collection of the information – as marks in form lists;
(2) The processing of this data – according to a given list of instructions;
(3) The final presentation of the results – as a form order.
These are the three basic stages when using a computer:

input – processing – output

input
In order to carry out such a task on a computer the machine must be capable of accepting the data as collected in (1), and also of accepting the list of instructions (or PROGRAM), needed to process this data. This is normally done through some form of INPUT device such as the CARD READER shown in figure 2.

At some stage all the information, both data and instructions, will have to be prepared on punched cards. Each card usually holds one piece of data or one instruction.

store
As each piece of information is accepted, the computer needs to STORE it away so that it may be available for use, as and when it is required. The store is the heart of the computer and can be thought of as a set of individual pigeon holes, each of which is capable of holding a single number or a single instruction. The pigeon holes are arranged in such a way that each can easily be identified, so that its contents may be examined or exchanged. Each will normally be given an address or label such as A, B, C . . . , or S1, S2, S3, . . . etc.

figure 2   A card reader

Every new number or instruction fed into the input device will be 'loaded' into a designated location, replacing the previous contents of that particular location. It will remain there until something is done to replace it.

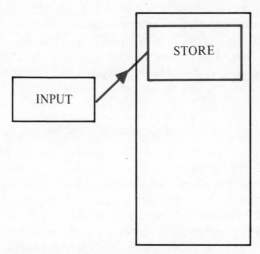

*figure 3   Passage of DATA and INSTRUCTIONS
from INPUT to STORE*

arithmetic unit

If the instructions in the program contain various arithmetic calculations the relevant data will have to be transferred, or rather COPIED, from the store to the part of the computer known as the ARITHMETIC UNIT. Here the actual arithmetic is carried out before returning the result of the calculation to another designated location in the store.

It is important to remember that only the location designated for the result of the particular calculation will have its contents changed.

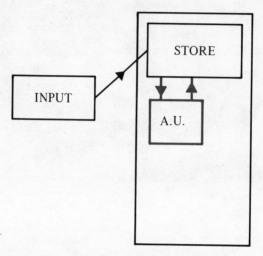

figure 4   Passage of DATA from STORE to
ARITHMETIC UNIT and vice versa

output

When the various calculations have been carried out, there must be some means of getting the results of the problems out of the computer in a meaningful form. This is done through some form of OUTPUT device, which in many cases now is a high speed LINE PRINTER capable of typing several hundred lines of print every minute.

control unit

Perhaps the most important part of the computer is the CONTROL UNIT. This part coordinates the working of all the other parts, and decides when each will come into action. The control unit looks at each instruction in turn and activates the relevant part or parts of the machine. In the case of an arithmetic instruction, it would cause the numbers involved to be passed from the store to the arithmetic unit, the operation to be carried out, and the result of the calculation to be returned to the store. In the case of an Input or Output instruction, it would cause the card reader to read the next card or the line printer to print out the relevant information.

figure 5   A line printer

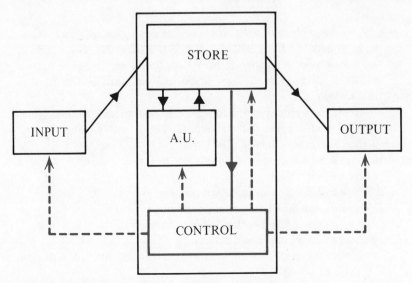

figure 6   Effect of CONTROL UNIT on other parts

other hardware

The main parts of the computer, consisting of the STORE, ARITHMETIC UNIT, and CONTROL UNIT are usually housed in a single box and are known as the CENTRAL PROCESSING UNIT (C.P.U.). The input and output devices, which are merely sophisticated ways of speeding up the passage of information into and out of the computer, are usually termed PERIPHERAL EQUIPMENT. All this is usually called hardware.

In addition to the above, there is normally a small typewriter CONSOLE linked to the C.P.U. Through this an operator can have limited access to the computer, and it is used to start and stop a particular job, or to check faults on the machine. Extra storage, in the form of interchangeable TAPE DECKS or DISC PACKS linked to the C.P.U., is also available. This is a way of giving fast access to large amounts of information without having to feed it into the computer each time, via the card reader.

The main object of this book is to show the type of instruction that the computer can understand and act on. It will also show how the computer uses these to carry out simple arithmetic calculations and to manipulate the data held in its store.

By showing how various problems can be broken down into separate stages, and how the allowable instructions can be combined, we shall see how the computer is able to carry out the more complex calculations necessary to solve these problems.

It will be useful to have a small model of a computer to illustrate various ideas. Some suggestions occur in the project which follows. This model computer will be developed as the book progresses, and will cater for each new idea as it arises.

project   the matchbox computer – storage

1   Make a model of the STORAGE UNIT of a computer. This should be capable of holding numbers in the various locations. Give each location a different label. You will need at least twelve locations for the work that follows.

It will be necessary to incorporate in your model, sections which can be used for the INPUT and OUTPUT, and possibly an area designated as the ARITHMETIC UNIT.

Here are some suggestions you might like to try:
  (a) a set of matchboxes;
  (b) a piece of cloth with individual pockets sewn on;
  (c) a cardboard box with several partitions;
  (d) a set of small tins (possibly tobacco) stuck onto a board;
  (e) a set of hooks on a board;
  (f) a piece of card, marked in large squares, and covered with clear Fablon or Perspex;
  (g) a tray with several partitions;

(h) a set of interlocking drawer units (as sold for screws and nails).

The data can be written on slips of paper or card, which can be put in the tray, drawer, or box, or written with felt pen on Perspex.

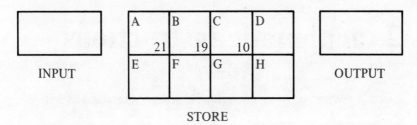

INPUT

STORE

OUTPUT

2 Enter three numbers 21, 19, 10, into the three locations as shown, and write a set of instructions telling an operator, using your computer, how to carry out the calculation (21 + 19) x 10. The result of the calculation should be put into a fourth location. Remember that in your instructions you will only be able to tell the operator where to look for the numbers, and not what the numbers actually are.

3 Enter three or four numbers into the store of your computer and write a set of instructions telling an operator how to carry out some calculation. Give a friend your list of instructions and your computer, with the numbers already in the appropriate locations. Ask him to find out what calculation the list of instructions was performing. Was he right?

2 arithmetic instructions

problem   What computer instruction is necessary to add 123 and 456?

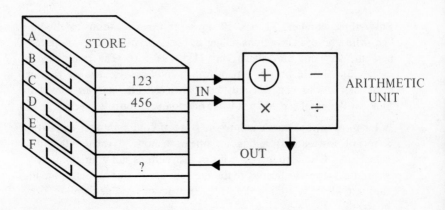

figure 1

Let us suppose, as shown in figure 1, that the numbers 123 and 456 are in locations A and B of the computer, and that we wish to add them together.

Since the computer is unable to deal with numbers directly and can only manipulate the contents of its storage locations, we want an instruction which tells the computer to add the number in location A to the number in location B. In addition the computer needs to know what to do with the result of the calculation. Let us assume that the result is to be placed in location E.

If you were to carry out this calculation on your matchbox computer, three separate instructions might be involved, namely:

(1) Copy the contents of location A into the arithmetic unit − 123
(2) Add the contents of location B into the arithmetic unit − 123 + 456
(3) Transfer the result from the arithmetic unit to location E − 579

Even in these three instructions we are assuming that the arithmetic unit is capable of doing the actual addition As a result a considerable amount will be implied by any instruction that we use.

One possible form of the instruction might be:

ADD the contents of location A to the contents of location B, and place the RESULT in location E.

Since the computer needs to know the location into which the result of the calculation is to be placed before it actually carries out the calculation, it is customary to designate this location first, for example:

REPLACE the contents of location E, by the contents of location A ADDED to the contents of location B.

Can you invent a form of shorthand to describe this instruction?

Some possibilities are: REPLACE location E by location A + location B

or REPLACE E by A + B

or E ← A + B

where '←' means 'is replaced by' or 'becomes' or 'is set equal to'. Providing we really understand what is entailed in this instruction we may use this form of shorthand.

dry check

It is very helpful to make a table showing the contents of each location before and after acting upon an instruction.

The effect of the instruction 'C ← A + B' is shown below, when the original contents of locations A, B, and C, were 73, 29, and 42 respectively:

	LOCATION A	LOCATION B	LOCATION C
INITIAL CONTENTS	73	29	42
INSTRUCTION C ← A + B			
NEW CONTENTS	73	29	102

This process is particularly useful when several instructions are involved, and is known as a DRY CHECK.

note: The contents of location A and location B remain unchanged, and ONLY the contents of the location named on the left hand side of the instruction, in this case location C, actually change.

3-address instruction

The instruction 'C ← A + B' is called a '3-ADDRESS INSTRUCTION', since the addresses, or labels, of three locations are involved, namely A, B, and C.

Other permissible 3-address instructions would be of the form:

$$D \leftarrow A - B$$
$$C \leftarrow A \times B$$
$$\text{or} \quad B \leftarrow C \div D$$

Table 1 shows how the contents of each location change as a result of the particular set of instructions. We shall see in a later chapter how the

table 1

	LOCATION A	LOCATION B	LOCATION C	LOCATION D
INITIAL CONTENTS	25	13	9	41
$A \leftarrow B + D$	54	13	9	41
$D \leftarrow A \div C$	54	13	9	6
$C \leftarrow B \times D$	54	13	78	6
$B \leftarrow C - A$	54	24	78	6

computer performs the operations of $\times$ and $\div$, by breaking them down into a set of simpler operations based on addition. For the present we must accept that the computer is capable of carrying out the $\times$ and $\div$, when it meets an instruction like '$C \leftarrow B \times D$' or '$D \leftarrow A \div C$'.

exercise 1

1 Copy and complete the following tables and show the contents of each location after carrying out the particular instruction.

(a)

	LOCATION A	LOCATION B	LOCATION C	LOCATION D
INITIAL CONTENTS	12	6	2	7
$C \leftarrow B + D$	12	6	13	7
$A \leftarrow B \times D$	?	?	13	7
$D \leftarrow A - B$	?	?	13	?
$C \leftarrow D \div B$	?	?	?	?

(b)

	LOCATION A	LOCATION B	LOCATION C	LOCATION D
INITIAL CONTENTS	19	5	21	42
$B \leftarrow C - A$				
$A \leftarrow D \div C$				
$C \leftarrow A + B$				
$D \leftarrow A \times B$				

2 If locations A, B, C, and D contain the numbers 10, 12, 26, and 3, what are the final contents of location G, as a result of carrying out each of the following lists of instructions?

(a) E ← C + D   (b) E ← A + B   (c) E ← C + B   (d) E ← A x A
     F ← A x B   F ← E + C   F ← A + E   F ← D + D
     G ← F + E   G ← F + D   G ← F ÷ D   G ← E x F

In each case do a dry check showing the successive contents of the locations.

3 In the following instructions state which are (i) meaningful, (ii) valid 3-address instructions.

(a) B ← A x A   (e) P ← Q ÷ R
(b) A + B ← C   (f) B + C → A
(c) D ← A − D   (g) A + B ← C + D
(d) A ← B + C − D   (h) A ← A x B + C

The same address can in fact occur in more than one place in an instruction as was seen in 2(d) above.

What is the effect of these instructions?

$$B ← A + A$$
$$C ← A − A$$

The first instruction tells the computer to add the contents of location A to itself and place the result into location B. In the second the effect is to put a nought into location C. Why?

Another useful instruction is

$$A ← A + B$$

Its effect is to add the contents of locations A and B and to put the result back into location A. This is shown in the table below.

	LOCATION A	LOCATION B
INITIAL CONTENTS	4	6
A ← A + B		
NEW CONTENTS	10	6

It is also possible to have the address of a location occurring in all three places, as the next instruction shows

$$A ← A + A$$

This instruction tells the computer to add the contents of location A to itself and to put the result back into location A − in other words it doubles the contents of location A.

What is the effect of these instructions?

$$A ← A x A$$
$$C ← C ÷ C$$
$$B ← B − B$$

Under what circumstances would the second be inadmissable?

economy of storage

It is considerably more economical to put the result of a calculation back into a location previously used, provided its original contents are not required later on in the calculation. As the storage in a computer is always limited it is good practice to get into the habit of doing this.

In 2(b) in the last exercise, seven locations were needed to carry out the addition of the numbers 10, 12, 26, and 3. An alternative version using only four locations is shown in table 2. Location A is used for the total.

	LOCATION A	LOCATION B	LOCATION C	LOCATION D
INITIAL CONTENTS	10	12	26	3
A ← A + B	22	12	26	3
A ← A + C	48	12	26	3
A ← A + D	51	12	26	3

Would the following list of instructions have the same effect?

$$A \leftarrow C + D$$
$$A \leftarrow A + B$$
$$A \leftarrow A + A$$

output instructions

Any list of instructions telling a computer how to carry out a calculation is known as a PROGRAM. In every program it is necessary to have an instruction which tells the computer to pass the final results of the calculation to the output device. Since a computer may have many thousand storage locations it would be impossible to guess where the result might be found.

An output instruction such as

OUTPUT from location A　or　OUTPUT A

will cause a copy of the contents of location A to be sent to the output device, where it might be printed onto paper. The computer operator can then collect the results printed by the line printer when it is convenient.

exercise 2

1　Locations A, B, C, and D contain the numbers 9, 7, 5, and 3 respectively. What is the result of carrying out each of the following programs?

(a) A ← A + B　　(b) A ← B x B　　(c) A ← A ÷ A　　(d) C ← A + B
　　C ← C + D　　　　A ← A x B　　　　OUTPUT A　　　　A ← A − B
　　A ← A ÷ C　　　　A ← A + B　　　　A ← A + A　　　　C ← A x B
　　OUTPUT A　　　　OUTPUT A　　　　OUTPUT A　　　　OUTPUT C

2   Rewrite the lists of instructions in exercise 1, question 2, using as few locations as possible and incorporating an output instruction for the final result.

3   If locations A, B, C and D contain the numbers 65, 110, 235, and 250 respectively, write programs to calculate and output the results of:

(a) $(65 + 110) \times 250$ (b) $65 + 110 \times 250$
(c) $(110 + 250) \div (65 + 235)$

project   the matchbox computer − output
1   Make sure your computer has an area which can be used for the OUTPUT device. This can be an extra tray, drawer or box etc. Carry out the following program on your computer using the numbers as shown in the diagram. You should finish with two numbers in the OUTPUT.

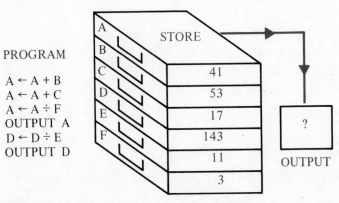

PROGRAM

A ← A + B
A ← A + C
A ← A ÷ F
OUTPUT A
D ← D ÷ E
OUTPUT D

STORE

41
53
17
143
11
3

?

OUTPUT

figure 2

2   Choose one of the five problems below. Decide which locations of your computer are going to hold the numbers in the problem. Write a list of instructions telling an operator, using your computer, how to carry out the calculation. Do not forget the necessary output instruction for the result.

(a) $\dfrac{14 \times 36 \times 45}{27 + 19}$ (b) $\dfrac{(14 \times 36) + 45}{27 + 19}$ (c) $(14 \times 36) + (45 \times 19) + 27$

(d) $(14 \times 36) + (45 \div 27) + 19$ (e) $(14 \times 36) + (45 \div 19) + 27$

Give your program and your computer with the numbers already stored in it to a friend. Ask him to carry out the instructions. When he has finished collect the result from the output device. Did he get the correct result? Ask him to decide which of the five problems you chose.

3 Make up a problem of your own. Choose some simple numbers and write a program to carry out the required calculation. Load the numbers into the store of your computer and ask a friend to carry out the program. Ask him to try and work out what your original calculation was.

3 flow charts

You had probably drawn a flow chart, or flow diagram as they are sometimes called, before you started on this book. In case you have not, a FLOW CHART is a diagram which describes how a particular situation is broken down into separate stages. In it the individual instructions are joined by arrowed lines which indicate the order in which the diagram is to be read. Because of its possible complexity, a flow chart must have a definite START point, and also a definite STOP, so that the reader knows where to begin and when he has finished.

Two flow charts are shown in figures 1 and 2. What does each achieve?

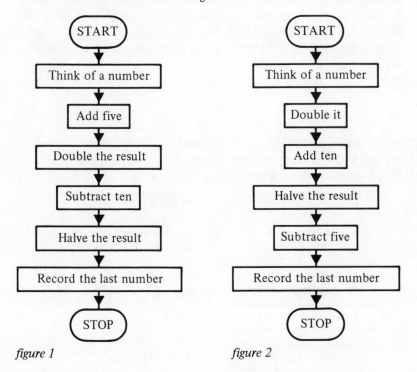

figure 1   figure 2

Did you discover that each flow chart produced the same result?

Note that even though the two flow charts appear to be very different, provided you started with the same number in both, the effect of each is the same.

It is often the case that a given problem can be done in more than one way. You should not be surprised if your flow chart for a particular problem does not always look exactly the same as anyone else's.

note Each instruction in the flow chart has been put in a rectangular box,

whereas the START and STOP points are entered in boxes with curved ends.

Later on, the flow chart will contain decisions or questions and these will require a third type of box. This is simply to make the reading quicker and easier to follow.

It is very often helpful to use a flow chart to describe the various stages in carrying out a calculation before attempting to write the detailed shorthand instructions in the program. In the problems that follow it is the order of the steps in the calculation that is important.

problem Use a computer to find the area of a circle for a given radius, r.

In this situation we shall need to know the formula which gives us the area of the circle, namely πr^2. A flow chart for this problem is shown in figure 3 with a possible program alongside. The computer will need to be given a suitable approximation to π e.g. 3·1416.

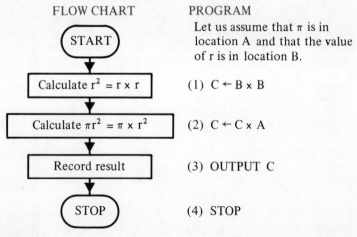

figure 3

problem   Use a computer to find the amount of metal used in making an open cylindrical tin, given that the radius of the base is r and its height is h.

In this problem the formula for the required area is $\pi r^2 + 2\pi rh$, πr^2 being the area of the circular base, and $2\pi rh$ being the area of the curved sides.

There are a number of possible ways of doing this calculation. One is shown in figure 4, outlined in the flow chart and carried out with a dry

FLOW CHART

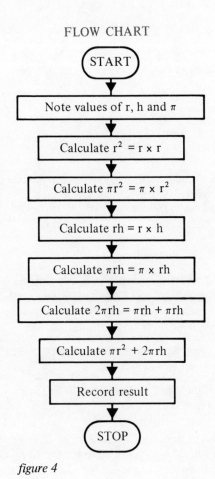

figure 4

check in the program. Note that it is simpler to calculate $2\pi rh$ by adding πrh to itself, rather than multiplying πrh by 2, which would necessitate having a location in the computer to hold the number 2.

Let us assume that π, r, and h are held in locations, A, B, and C respectively, with π to be 3·1416.

PROGRAM	DRY CHECK					
	LOCATION A	LOCATION B	LOCATION C	LOCATION D	LOCATION E	LOCATION F
START	π	r	h	?	?	?
(1) D ← B x B	π	r	h	r^2		
(2) D ← D x A	π	r	h	πr^2		
(3) E ← B x C	π	r	h	πr^2	rh	
(4) E ← E x A	π	r	h	πr^2	πrh	
(5) E ← E + E	π	r	h	πr^2	$2\pi rh$	
(6) F ← D + E	π	r	h	πr^2	$2\pi rh$	$\pi r^2 + 2\pi rh$
(7) OUTPUT F STOP						$\pi r^2 + 2\pi rh$

program 1

Can you find a different way of doing this calculation?

exercise 3

1 Draw flow charts to show how you would evaluate the following expressions:

(a) Volume of cylinder $\pi r^2 h$

(b) Height of cylinder $V \div \pi r^2$

(c) Sum of two squares $a^2 + b^2$

(d) Value of the determinant $\begin{vmatrix} a & b \\ c & d \end{vmatrix}$

(e) Surface area of a sphere $4\pi r^2$

(f) Volume of a sphere $\frac{4}{3}\pi r^3$

(g) Simple interest $\dfrac{P \times R \times T}{100}$

(h) Distance $\frac{1}{2}(u + v)t$

2 In each of the above questions write a program to carry out the required calculation. You will first need to decide which locations are to hold which pieces of data. Try and use as few locations in your program as possible.

3 Draw a flow chart to show how you would evaluate x^{13}. Write a program to carry out this calculation.

4 Look at your flow chart and program in question 3 very carefully. Could you have used
(a) fewer storage locations
or (b) fewer instructions?
What is the least number of instructions needed to calculate x^{13}? What is the least number of locations required in the calculation of x^{13}?

types of data

In most questions in the previous exercise there are two types of data. There are numbers such as π, 4, 3, 100, 2, which are known as fixed data, and quantities such as r, h, P, R, T, etc., whose value will depend on the particular situation. The latter are called variable data.

input instruction

Up till now we have assumed that any piece of data needed in a problem was in a particular location in the computer before we started on the calculation. The question naturally arises — how did it get there?

The computer in fact enters the data into a designated location when it is asked to carry out an INPUT instruction such as

Enter 26 into location A

or in shorthand INPUT 26 to A

We shall see in the next chapter, one way in which the computer is able to execute this instruction.

project   the matchbox computer – input instruction
Carry out program 2 using your matchbox computer. What result do you finish up with in the output device?

PROGRAM

START

(1) INPUT 17 to A

(2) INPUT 43 to B

(3) C ← A + B

(4) INPUT 25 to A

(5) INPUT 15 to B

(6) D ← A + B

(7) E ← C ÷ D

(8) OUTPUT E

STOP

program 2

drawing flow charts

When outlining how a problem is going to be tackled it is well worth remembering the three basic stages on the computer — INPUT, CALCU-LATION, OUTPUT. The first section of the flow chart should cover the entering of both types of data, the middle section will deal with the details of the calculation, and there will be a final section where the required results will be output.

problem   Find the average of three numbers.

A flow chart describing how this problem might be tackled on a computer is shown in figure 5. One instruction box has been left out. Which is it?

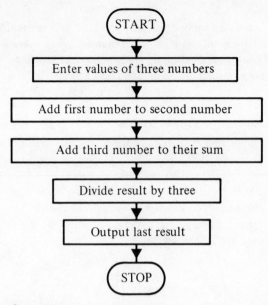

figure 5

Whilst this flow chart certainly describes how a human would find the average of three numbers, the computer would be faced with a problem in the fourth instruction — 'Divide result by three'. The computer needs a 3 for the division, but at no stage has a 3 been entered into a storage location.

The necessary missing instruction box, situated after the START would be

$$\boxed{\text{Enter a 3}}$$

Write out a correct program for this problem.

A similar situation arises if we wish to find the volume of a sphere using the formula $\frac{4}{3}\pi r^3$. Here the fixed data might appear a little more obvious, namely π, 4, and 3. Each of these would have to be entered into the computer's store in addition to the value of the variable, r.

A correct flow chart is shown in figure 6.

stages in writing a program

If we now wish to write a program to carry out this calculation, we must first decide which locations are to hold the data, and at the same time it is helpful to designate certain locations to be used for the working and the result. In this case we might use the following arrangement, taking π to be 3·1416. Location $A - \pi$, $B - 4$, $C - 3$, $D - r$, $E -$ workspace, $F -$ result.

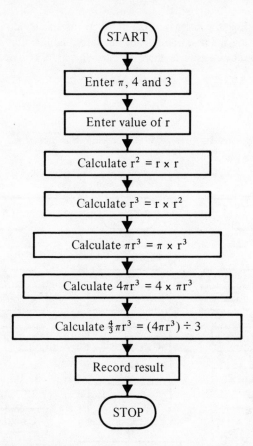

figure 6

PROGRAM   DRY CHECK

	A	B	C	D	E	F	OUTPUT
START	?	?	?	?	?	?	
(1) INPUT π to A	π						
(2) INPUT 4 to B		4					
(3) INPUT 3 to C			3				
(4) INPUT r to D				r			
(5) E ← D × D					r^2		
(6) E ← E × D					r^3		
(7) E ← E × A					πr^3		
(8) E ← E × B					$4\pi r^3$		
(9) F ← E ÷ C						$\frac{4}{3}\pi r^3$	
(10) OUTPUT F							$\frac{4}{3}\pi r^3$ from F
STOP							

program 3

While writing the actual program it is useful to do a dry-check alongside, noting the new contents of the particular location after carrying out each instruction (see program 3 and dry check).

exercise 4

1 Write programs for each of the calculations described in the flow charts in figure 7.

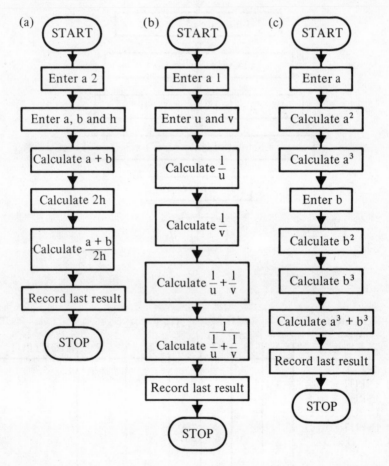

(a) START
Enter a 2
Enter a, b and h
Calculate a + b
Calculate 2h
Calculate $\dfrac{a + b}{2h}$
Record last result
STOP

(b) START
Enter a 1
Enter u and v
Calculate $\dfrac{1}{u}$
Calculate $\dfrac{1}{v}$
Calculate $\dfrac{1}{u} + \dfrac{1}{v}$
Calculate $\dfrac{1}{\dfrac{1}{u} + \dfrac{1}{v}}$
Record last result
STOP

(c) START
Enter a
Calculate a^2
Calculate a^3
Enter b
Calculate b^2
Calculate b^3
Calculate $a^3 + b^3$
Record last result
STOP

figure 7

2 Draw flow charts describing the calculation of the problem carried out in each of programs 4 to 6.

```
         START              START                START
(1) INPUT 2 to A     (1) INPUT 3 to A      (1) INPUT 2 to A
(2) INPUT b to B     (2) INPUT π to B      (2) INPUT u to B
(3) INPUT h to C     (3) INPUT r to C      (3) INPUT t to C
(4) D ← B x C        (4) INPUT h to D      (4) INPUT f to D
(5) D ← D ÷ A        (5) E ← C x C         (5) E ← B x C
(6) OUTPUT D         (6) E ← E x D         (6) F ← C x C
    STOP             (7) E ← E x D         (7) F ← F x D
                     (8) E ← E ÷ 3         (8) F ← F ÷ 2
                     (9) OUTPUT E          (9) E ← E + F
                         STOP             (10) OUTPUT E
                                               STOP

   program 4           program 5             program 6
```

graph plotting

In many cases when plotting graphs we have to evaluate a particular expression for a set of different values of x. For each x the steps of the calculation are basically the same.

problem   Evaluate $\dfrac{1}{1 + x^2}$ for a given value of x.

Often it helps to draw an outline flow chart describing the general approach to the problem, before drawing a detailed one showing the exact steps of the calculation. For this problem they might be as shown in figure 8.

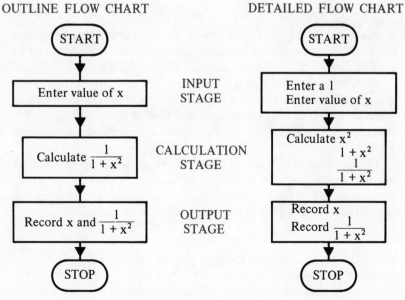

figure 8

It was probably only after drawing the outline flow chart that the need for a ONE became apparent. Note once again the three basic stages in the problem.

In this situation we shall need three locations, one for the 1, one for the x, and one for the result. A program with a dry check is shown below.

PROGRAM	DRY CHECK			
	A	B	C	OUTPUT
START	?	?	?	
(1) INPUT 1 to A	1			
(2) INPUT x to B		x		
(3) $C \leftarrow B \times B$			x^2	
(4) $C \leftarrow C + A$			$1 + x^2$	
(5) $C \leftarrow A \div C$			$\dfrac{1}{1 + x^2}$	
(6) OUTPUT B				x from B
(7) OUTPUT C				$\dfrac{1}{1 + x^2}$ from C
STOP				

program 7

How could we amend the flow chart to evaluate $\dfrac{1}{1 + x^2}$ for several values of x?

stored programs

It was stated in chapter 1, that one of the main differences between a computer and a calculating machine, was the ability of the computer to store its own list of instructions and carry them out automatically.

Each instruction in the program will be stored in a separate location, in sequence. When the computer is given the address of the first instruction it will act on each in turn and follow them through successively until it is told to stop. This means that the final instruction in any program must be a STOP instruction.

Up till now when using your matchbox computer the list of instructions has been on a piece of paper. We can simulate the stored program facility of the computer by writing each instruction on a separate piece of paper or card, and entering them into part of the storage unit.

project   the matchbox computer – program store

Designate part of your matchbox computer store to hold instructions. If you have not enough locations for this, then add a section which can be used to store the program.

It is a good idea to label the individual locations which are going to hold the instructions as (1), (2), (3), ... etc. to distinguish them from the data locations A, B, C, D, etc.

1   A set of eight instructions is stored in the computer as shown in figure 9. The first instruction is in LOCATION (1). Follow the instructions through in sequence and find out what the program does.

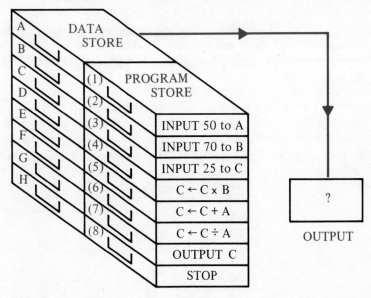

figure 9

Why is is necessary to have a STOP as the last instruction of the program?

2   Make up a program of your own. Write the individual instructions on separate cards. Put the instructions in sequence into the part of your computer designated as the program store. Ask a friend to carry out the instructions as they are written. Remember to tell him in which location the first instruction is. Remember to make the final instruction in your program a STOP instruction.

4 jump instructions

In the last chapter it was shown how the computer could evaluate the expression $1/(1 + x^2)$ for a given value of x. When plotting a graph of this function it would be necessary to evaluate $1/(1 + x^2)$ for a set of different values of x.

Since, for each x, the steps of the calculation are basically the same, this repetitive situation is one ideally suited to the computer.

The problem was posed as to how the flow chart might be amended to cater for several values of x. This can be done quite easily by adding a return loop after the output instruction. This loop would incorporate an instruction such as 'Enter next value of x', and re-enter the flow chart just before the instruction 'Calculate x^2' (see figure 1).

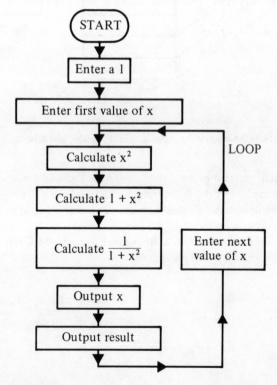

figure 1

Two snags arise with this modified flow chart:

  (1) Where would we find the next value of x?

  (2) How long would we continue to go round the LOOP?

If we were ourselves following through the flow chart or program, each time we went round the loop and came to the instruction 'Enter next value of x' we would ask ourselves the question 'Are there any more x's?' This would require a decision on our part, with the possible answers YES or NO. We shall see in chapter 5 how the computer can make such a decision.

We can modify the program, for evaluating $1/(1 + x^2)$ (see page 24), in a similar way by changing the second instruction to read 'Input next x to B' and by adding the JUMP INSTRUCTION, 'GO TO (2)' after the second output instruction. (See program 1.)

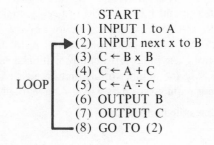

         START
      (1) INPUT 1 to A
      (2) INPUT next x to B
      (3) C ← B x B
      (4) C ← A + C
LOOP (5) C ← A ÷ C
      (6) OUTPUT B
      (7) OUTPUT C
      (8) GO TO (2)

program 1

jump instruction

Here the use of an instruction, which tells the computer to go back to a previous instruction, enables the idea of a LOOP to be incorporated into the program. The instruction

GO TO (2)

is known as an UNCONDITIONAL JUMP INSTRUCTION, since having carried out instruction (7) the computer will always go back to instruction (2).

The question also arises 'Where does the computer find the next value of x?'

On meeting an instruction such as 'INPUT next data to B', the control unit in the computer will cause the next number waiting in the input device to be 'loaded' into the designated location B.

project   the matchbox computer – input
Make sure your matchbox computer has an area which can be used for the input device. This can be an extra tray, drawer or box etc.

1 Write three numbers on separate cards and put them into the input device and carry out program 2 using your computer.

> START
> (1) INPUT next data to A
> (2) INPUT next data to B
> (3) INPUT next data to C
> (4) D ← A + B
> (5) D ← D + C
> (6) OUTPUT D
> STOP
>
> *program 2*

Is the SUM of the three numbers now in the output?

If we wanted to add together a list of twenty numbers, it would be very tedious to write out twenty instructions of the form 'Input next data to store'. Also we should need several instructions adding the numbers. By using the jump instruction mentioned earlier we can cut down considerably the number of instructions needed. The number of storage locations required for data will also be reduced.

Program 3 uses the jump instruction.

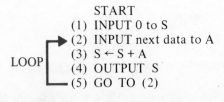

> START
> (1) INPUT 0 to S
> (2) INPUT next data to A
> LOOP (3) S ← S + A
> (4) OUTPUT S
> (5) GO TO (2)

program 3

2 Put several numbers into the input device of your computer and carry out program 3. This time you will have several results in your output device. Is the final number in output the sum of the original numbers?

Would you get the same results in output if you had put the numbers into the input in a different order?

3 Carry out the program again with a different set of numbers but this time leave out instruction (1).

Why is this instruction so important?

problem Write a program to calculate a batsman's successive averages during the season.

In this situation, after each innings we need to calculate the total number of runs scored by the batsman. These results will then be divided by the total number of innings completed up to that particular time. A record of the number of innings must be kept.

It is useful to designate two locations, one to hold the number of runs scored (SUM), and the second to hold the number of innings (COUNTER). As you saw in the last program these must initially be set to zero, so that the data is not added to the numbers left from a previous problem.

A flow chart describing how this problem might be tackled is shown in figure 2.

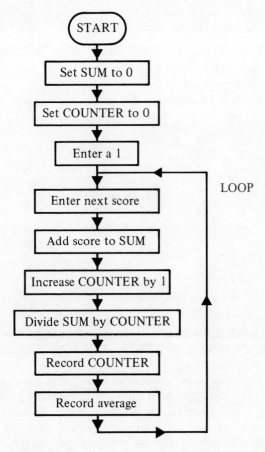

figure 2

note: The ONE is necessary in order to be able to increase the COUNTER by 1, each time a new score is entered. Once again the LOOP enables us to use the same set of instructions for each successive calculation. The scores for each innings would be held in the INPUT until called for by the program. To deal with 'NOT-OUTS' the flow chart would involve the use of a question.

modified input instructions

In order to distinguish between the fixed data in a problem and the variable data, we can develop the shorthand already used, for our INPUT instructions.

Where we have fixed data, and an input instruction like 'INPUT 1 to B', or 'INPUT 0 to C', we will allow the use of an instruction such as

$$B \leftarrow 1 \quad \text{or} \quad C \leftarrow 0$$

The '$\leftarrow$' again means 'is replaced by' or 'is set equal to'.

On the other hand for variable data, and an input instruction such as 'Input next number to B', we can shorten this to

INPUT B

An instruction like this will imply that the data to be loaded into LOCATION B will be found in the input device.

Program 4 uses this notation for the successive averages problem.

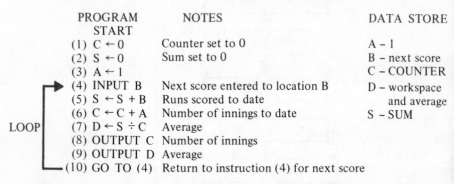

PROGRAM START	NOTES	DATA STORE
(1) $C \leftarrow 0$	Counter set to 0	A – 1
(2) $S \leftarrow 0$	Sum set to 0	B – next score
(3) $A \leftarrow 1$		C – COUNTER
(4) INPUT B	Next score entered to location B	D – workspace
(5) $S \leftarrow S + B$	Runs scored to date	and average
(6) $C \leftarrow C + A$	Number of innings to date	S – SUM
(7) $D \leftarrow S \div C$	Average	
(8) OUTPUT C	Number of innings	
(9) OUTPUT D	Average	
(10) GO TO (4)	Return to instruction (4) for next score	

LOOP (brackets instructions 4 to 10)

program 4

In a given problem, it would be perfectly possible to have all the required data held in the input device. This could then be called for, using the appropriate input instruction.

However the development of the modern computer has made it possible for certain instructions, like '$B \leftarrow 1$', to be executed automatically.

For every instruction used in this book we have assumed that the computer could carry out a number of fairly complex actions. We saw, at the beginning of chapter 2, that the addition of two numbers necessitated each being moved to the arithmetic unit, the addition performed, and the result placed in an appropriate storage location. To carry out the addition, the binary (0's and 1's) representations of each number have to be combined in a particular way. Even this requires several instructions to cater for 'carrying' etc. Instructions at this level are known as the computer's MACHINE CODE.

In the early days of computing every program using addition would have consisted of these machine code instructions. Now, rather than doing this whenever the addition is required, it is possible to arrange for the computer to translate our instructions, such as 'A ← B + C', into the minute detail of the machine code.

The translation is done by a vast program called a COMPILER. This is supplied by each manufacturer in their particular computer's machine code. The compiler program is first fed into the computer and translates any program which follows into the type of instruction which the computer can actually execute.

The compiler could generate certain constants like 1, 2, or 0 and hence translate the instructions we have allowed such as 'B ← 1' or 'C ← 0'. Every compiler will have its own list of instructions which it is capable of translating. In theory one could design a compiler to translate almost any instruction, but it would then almost certainly fill the whole of the available store of the computer. If this were the case, there would be no room left for the data and instructions of the problem one wanted to solve!

The real difference between a true computer and a programmable calculator is the ability of the computer to amend or modify its own instructions.

exercise 5

1   In programs 5-10 the INPUT device holds the variables x, y, and z, in that order. Carry out a DRY CHECK on each program and find the algebraic expression that each program evaluates.

START	START	START
(1) INPUT A	(1) INPUT A	(1) INPUT A
(2) INPUT B	(2) INPUT B	(2) INPUT B
(3) C ← A − B	(3) INPUT C	(3) INPUT C
(4) D ← A × B	(4) A ← A + B	(4) D ← A × B
(5) C ← C ÷ D	(5) A ← A + C	(5) D ← D + C
(6) OUTPUT C	(6) OUTPUT A	(6) OUTPUT D
STOP	STOP	STOP
program 5	*program 6*	*program 7*

START	START	START
(1) INPUT A	(1) INPUT A	(1) D ← 3
(2) INPUT B	(2) INPUT B	(2) INPUT A
(3) INPUT C	(3) C ← A + B	(3) INPUT B
(4) D ← A × C	(4) D ← A × B	(4) INPUT C
(5) E ← B × C	(5) A ← A × A	(5) E ← A × B
(6) D ← D + E	(6) B ← B × B	(6) D ← D × E
(7) E ← A × B	(7) D ← D + A	(7) E ← A × C
(8) D ← D ÷ E	(8) D ← D + B	(8) F ← B × C
(9) OUTPUT D	(9) D ← D ÷ C	(9) E ← E + F
(10) STOP	(10) OUTPUT D	(10) D ← E ÷ D
	STOP	(11) OUTPUT D
		STOP
program 8	*program 9*	*program 10*

In which of these programs is the order of the data in the input device important?

2 Translate each of the flow charts in figure 3 into a program. Before you write the program decide which locations are going to hold the data, and which locations will be required for the working.

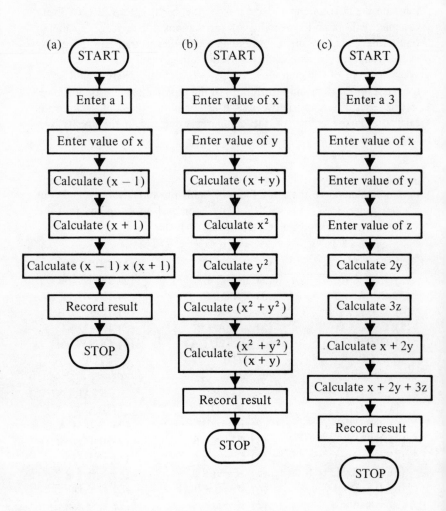

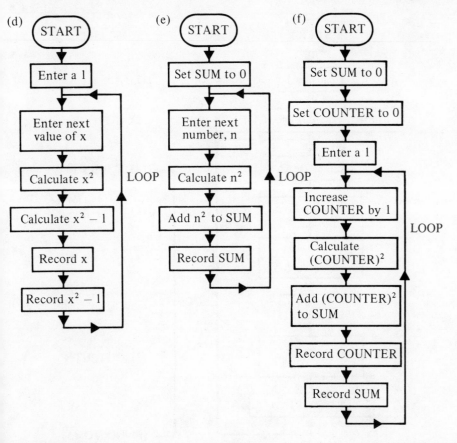

figure 3

counter

The idea of using a location to act as a COUNTER is a very important one in programming. In the program to evaluate $1/(1 + x^2)$ it is possible that the particular values of x would have been the set of whole numbers from -4 to $+4$.

If we designate a location, which initially is set to -4, to hold x, we could get the computer to work out each successive value of x by adding 1 to the previous value.

If location A held 1 and location B held x, the instruction 'B $\leftarrow$ B + A' would increase the contents of location B by 1. By incorporating a loop to repeat the calculation for the value of x, just worked out, the program would evaluate $1/(1 + x^2)$ for as many x's as desired.

The flow chart in figure 4 and program 11 show the modifications.

34

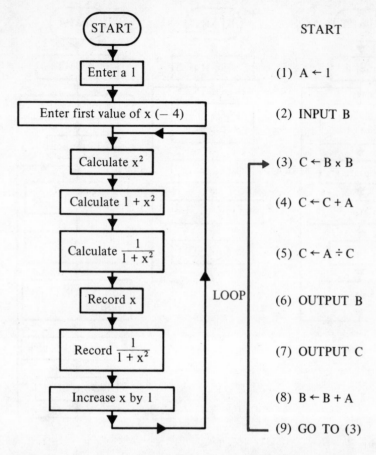

figure 4   program 11

note In this case the input device will contain the first value of x. When instruction (2) is carried out this value (−4) will be loaded into location B.

exercise 6

1   Draw flow charts to show how you would evaluate each of the following expressions.

    (a) x^3 for $x = 0, 1, 2, 3, \ldots$
    (b) $x^2 + 1$ for $x = -3, -2, -1, 0, 1, 2, \ldots$
    (c) $3x^2 + x$ for $x = -6, -5, -4, -3, -2, -1, 0, 1, 2, \ldots$
    (d) $5x^2 + 3x + 1$ for $x = 0, 2, 4, 6, \ldots$
    (e) $2 - x^2$ for $x = 0, 0 \cdot 5, 1 \cdot 0, 1 \cdot 5, 2 \cdot 0, 2 \cdot 5, \ldots$

2   Write programs to carry out each of the tasks outlined in the flow charts you have drawn for question 1.

     If you can, run your programs on a computer and use your results to plot the graphs of the various expressions.

3   Carry out a dry check on each of programs 12 to 17 and determine what each is doing. You may assume that the input device holds the number 0.

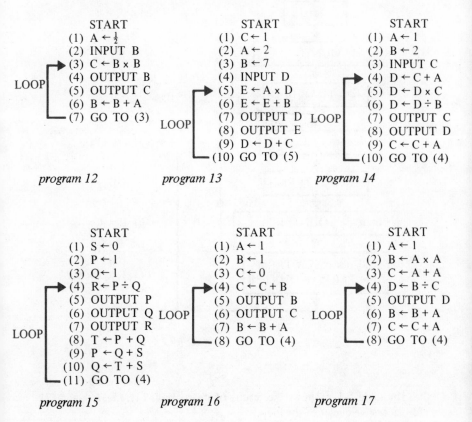

	START
	(1) A ← ½
	(2) INPUT B
LOOP	(3) C ← B x B
	(4) OUTPUT B
	(5) OUTPUT C
	(6) B ← B + A
	(7) GO TO (3)

program 12

	START
	(1) C ← 1
	(2) A ← 2
	(3) B ← 7
	(4) INPUT D
	(5) E ← A x D
	(6) E ← E + B
LOOP	(7) OUTPUT D
	(8) OUTPUT E
	(9) D ← D + C
	(10) GO TO (5)

program 13

	START
	(1) A ← 1
	(2) B ← 2
	(3) INPUT C
	(4) D ← C + A
	(5) D ← D x C
LOOP	(6) D ← D ÷ B
	(7) OUTPUT C
	(8) OUTPUT D
	(9) C ← C + A
	(10) GO TO (4)

program 14

	START
	(1) S ← 0
	(2) P ← 1
	(3) Q ← 1
	(4) R ← P ÷ Q
	(5) OUTPUT P
	(6) OUTPUT Q
	(7) OUTPUT R
LOOP	(8) T ← P + Q
	(9) P ← Q + S
	(10) Q ← T + S
	(11) GO TO (4)

program 15

	START
	(1) A ← 1
	(2) B ← 1
	(3) C ← 0
	(4) C ← C + B
	(5) OUTPUT B
LOOP	(6) OUTPUT C
	(7) B ← B + A
	(8) GO TO (4)

program 16

	START
	(1) A ← 1
	(2) B ← A x A
	(3) C ← A + A
	(4) D ← B ÷ C
	(5) OUTPUT D
LOOP	(6) B ← B + A
	(7) C ← C + A
	(8) GO TO (4)

program 17

problem   Write a program to calculate and print out n! for n = 1, 2, 3, . . . Here is a situation where we can calculate each new value of n! from the previous value. For example 4! = 4 x 3! and 5! = 5 x 4!

It is useful to designate one location to hold the current TERM or value of n!, and another to hold the value of n, which can be thought of as a COUNTER, indicating how many values of n! we have already calculated. In order to calculate the next value of n! we need to multiply each TERM, (n!) by the next value of the COUNTER, (n + 1). This is summarized in the flow chart in figure 5 and carried out in program 18.

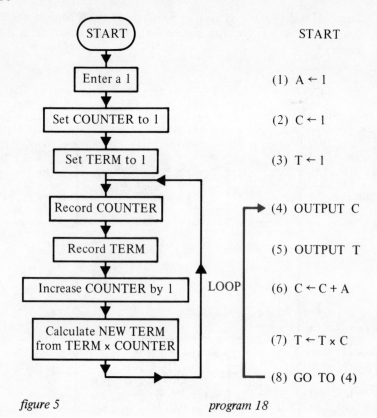

figure 5   *program 18*

The table below shows the results of instructions (4), (5), (6), and (7) for successive cycles of the loop.

		1st cycle	2nd cycle	3rd cycle	4th cycle
Instruction (4)	OUTPUT C	1	2	3	4
(5)	OUTPUT T	1!	2!	3!	4!
(6)	next COUNTER	1 + 1	2 + 1	3 + 1	4 + 1
(7)	new TERM	1 × 2	1 × 2 × 3	1 × 2 × 3 × 4	1 × 2 × 3 × 4 × 5 etc

problem   Write a program to calculate and print out the following SUMS.

$$1 + \tfrac{1}{2}$$
$$1 + \tfrac{1}{2} + \tfrac{1}{4}$$
$$1 + \tfrac{1}{2} + \tfrac{1}{4} + \tfrac{1}{8}$$
$$1 + \tfrac{1}{2} + \tfrac{1}{4} + \tfrac{1}{8} + \tfrac{1}{16} \text{ etc.}$$

In this situation we can calculate each successive SUM by adding the next TERM in the sequence $1, \tfrac{1}{2}, \tfrac{1}{4}, \tfrac{1}{8}, \tfrac{1}{16}, \tfrac{1}{32}, \ldots$ to the previous SUM.

Here it is a good idea to designate one location to hold the current TERM, and a second to hold the SUM. We shall also need a location containing 2, since each TERM may be obtained by dividing the previous TERM by 2. For example $\frac{1}{32} = \frac{1}{16} \div 2$.

The flow chart in figure 6 describes how the problem might be tackled.

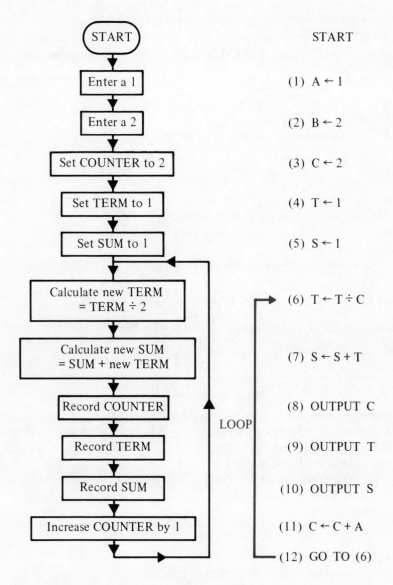

figure 6   *program 19*

note The addresses of the locations used in program 19, correspond where possible, to the role that the particular location plays in the problem – location T for TERM, location S for SUM, location C for COUNTER. In the program, the counter is being used to indicate how many terms are being summed at the particular stage. The three outputs are the number of terms C, the value of the current term T, and the value of the sum S. The first and second set of outputs would be $C = 2$, $T = \frac{1}{2}$, and $S = 1 + \frac{1}{2}$, and $C = 3$, $T = \frac{1}{4}$, $S = 1 + \frac{1}{2} + \frac{1}{4}$ respectively.

Make a table as in the previous problem, showing the results of instructions (6), (7), (8), (9), (10), and (11), for each successive cycle of the loop and check that the above results are correct.

exercise 7

1 Write programs to calculate and print out, when $n = 1, 2, 3, 4, \ldots$, the values of

(a) n^3 (b) 2^n (c) $\dfrac{1}{2^n}$ (d) $\dfrac{1}{n!}$

2 Write programs to calculate and print out the following SUMS.

(a) $1^3 + 2^3$
$1^3 + 2^3 + 3^3$
$1^3 + 2^3 + 3^3 + 4^3$ etc.

(b) $1 + \frac{1}{3}$
$1 + \frac{1}{3} + \frac{1}{9}$
$1 + \frac{1}{3} + \frac{1}{9} + \frac{1}{27}$ etc.

(c) $2 + 2^2$
$2 + 2^2 + 2^3$
$2 + 2^2 + 2^3 + 2^4$ etc.

(d) $1 + 3$
$1 + 3 + 5$
$1 + 3 + 5 + 7$ etc.

(e) $1 + \frac{1}{1!}$
$1 + \frac{1}{1!} + \frac{1}{2!}$
$1 + \frac{1}{1!} + \frac{1}{2!} + \frac{1}{3!}$ etc.

(f) $1 + \frac{1}{2}$
$1 + \frac{1}{2} + \frac{1}{3}$
$1 + \frac{1}{2} + \frac{1}{3} + \frac{1}{4}$ etc.

The idea of a COUNTER, and increasing the counter by one, together with that of designating certain locations to hold the current TERM and SUM, will crop up again and again. These are some of the main techniques and building blocks of programming.

Another important idea is that of replacing the contents of one location by the contents of a second. This occurred in exercise 6, question 3, program 15.

replacement instruction

The instruction 'P ← Q + S', where location S contains zero, in fact duplicates the contents of location Q in location P. Similarly the instruction 'P ← Q x A', where location A contained one, would have the same effect.

We will allow a simpler form of these instructions to eliminate the need to provide a location holding a zero or a one.

$$P \leftarrow Q$$

This instruction replaces the contents of location P with those of location Q, but at the same time leaves the contents of location Q unchanged. Program 14 of question 3 in exercise 6 was to calculate the various terms in the Fibonacci Sequence 1, 1, 2, 3, 5, 8, 13, . . . etc., and also the ratio of successive terms of that sequence $\frac{1}{1}, \frac{1}{2}, \frac{2}{3}, \frac{3}{5}, \frac{5}{8}, \ldots$ etc.

Since the calculation only requires finding the next term of the sequence by adding the previous two terms together, and then finding the ratio of the two terms, it is convenient to use the idea of the loop, having first replaced each pair of terms to be considered, by the next pair of terms.

The full flow chart and program using the modified replacement instruction, 'P ← Q', are shown in figure 7 and program 20.

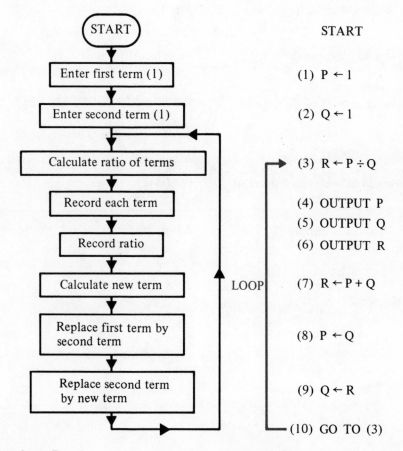

figure 7   *program 20*

Change instructions (1) and (2) in figure 7 to INPUT P and INPUT Q and place any two numbers in the INPUT. What effect does this have on the successive ratios?

So far the question 'How do we get out of a LOOP in a program?' has been left unanswered. This will be dealt with in the next chapter.

summary of allowable instructions
A summary of the allowable instructions met so far is shown below.

arithmetic instructions	$A \leftarrow B + C$
	$A \leftarrow B - C$
	$A \leftarrow B \times C$
	$A \leftarrow B \div C$
also those of the type	$A \leftarrow A \times A$

output instruction   OUTPUT A

input instructions   INPUT A — for variable data held in the input device.

                                   $A \leftarrow 1$ — for fixed data.

unconditional jump instruction   GO TO (3)

replacement instruction $A \leftarrow B$

5 conditional jump instructions

In the last chapter we saw how a list of instructions could be used over and over again by employing a loop and a JUMP INSTRUCTION.

This raised two questions which need to be answered:

(1) How does the computer know whether or not there is any more data for an 'Input next data' instruction?

(2) How does the computer know when to stop going round a loop, if it is calculating successive terms in a sequence, or the values for a graph?

The computer in fact has the ability to answer certain types of questions, provided they can be answered with a YES or a NO. The computer is able to make comparisons between two numbers and can answer a question such as 'Is this number less than that number?'

decision box

In a flow chart we can show a question by using a DECISION BOX such as

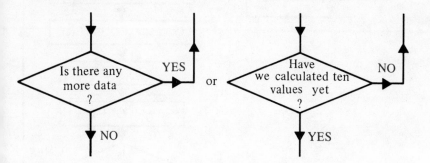

Each decision box will have two possible exits, one if the answer to the question is YES, and the other if the answer to the question is NO. Normally one of these exits will return to a previous point in the flow chart.

The diamond-shaped box is used to distinguish a question in the flow chart from an instruction.

problem   Use a computer to print out the first ten natural numbers and their cubes.

In this situation we could employ a COUNTER to indicate which cube was being printed out at the particular time. If we use a question of the form

Is the COUNTER < 10?

we would want to go back and calculate the next cube when the answer to the question was YES, but go on to a STOP instruction when the answer was NO.

A flow chart describing the details of the calculation, together with one giving first outline thoughts is shown in figure 1.

OUTLINE FLOW CHART   DETAILED FLOW CHART

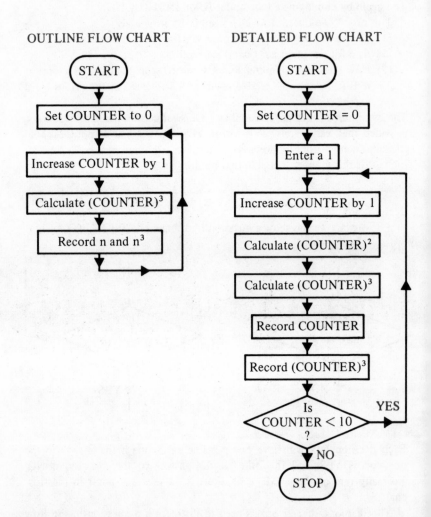

figure 1

conditional jump instruction

In the program carrying out this calculation, the instruction corresponding to the question in the decision box might be

$$\text{If the COUNTER is} < 10 \quad \text{GO TO Instruction (3)}$$

or in shorthand $\qquad$ IF $C < 10$ $\quad$ GO TO (3)

This is known as a CONDITIONAL JUMP INSTRUCTION, since the jump to instruction (3) will only take place if the condition has been satisfied.

The full program is given in program 1 with a dry check alongside.

	PROGRAM START	RESULT OF INSTRUCTIONS ON SUCCESSIVE LOOPS			
		1st LOOP	2nd LOOP	3rd LOOP	...10th LOOP
	(1) $A \leftarrow 1$				
	(2) $C \leftarrow 0$				
→	(3) $C \leftarrow C + A$	$0 + 1$	$1 + 1$	$2 + 1$	$9 + 1$
	(4) $D \leftarrow C \times C$	1^2	2^2	3^2	10^2
	(5) $D \leftarrow D \times C$	1^3	2^3	3^3	10^3
	(6) OUTPUT C	1	2	3	10
	(7) OUTPUT D	1	8	27	1000
	(8) IF $C < 10$ GO TO (3)	YES	YES	YES	... NO
	(9) STOP				

LOOP — YES (at left margin)

program 1

note Since the computer tends to work in terms of the storage locations rather than actual numbers, the instruction IF $C < 10$ GO TO (3) is normally written in the form

$$\text{IF } C < B \text{ GO TO (3)}$$

where location B has been loaded with the number 10.

multiplication

It was mentioned in chapter 2 that the arithmetic unit of the computer was only capable of performing addition and that at some stage the operations of $\times$ and $\div$ had to be broken down into a set of instructions based on addition.

The computer carries out multiplication by the process of repeated addition, for example the calculation 26×3 can be written as $26 + 26 + 26$.

By employing a COUNTER and asking the question 'Has counter reached 3 yet?' the computer can determine when to stop the repeated addition. Storage locations will be required for both the 26 (location N) and the 3 (location M), and the program will involve the conditional jump instruction:

$$\text{IF } C < M \text{ GO TO (3)}$$

The flow chart and program for this are shown in figure 2 and program 2.

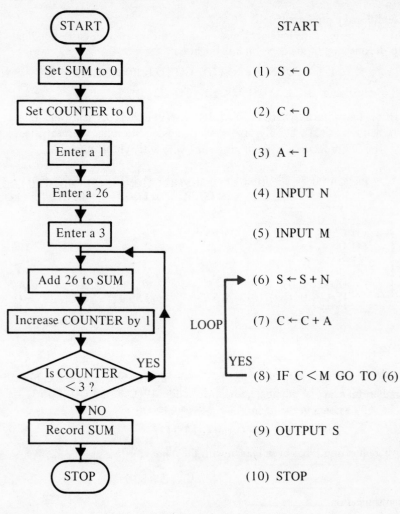

figure 2   *program 2*

note The input device will contain the numbers 26 and 3. These will be loaded into locations N and M as a result of instructions (4) and (5). The two problems which follow illustrate these ideas further.

problem Find the average of 30 given marks.
In this situation we can use a counter, and the question 'Is the counter < 30?' to determine whether all 30 marks have been entered. See figure 3 and program 3.

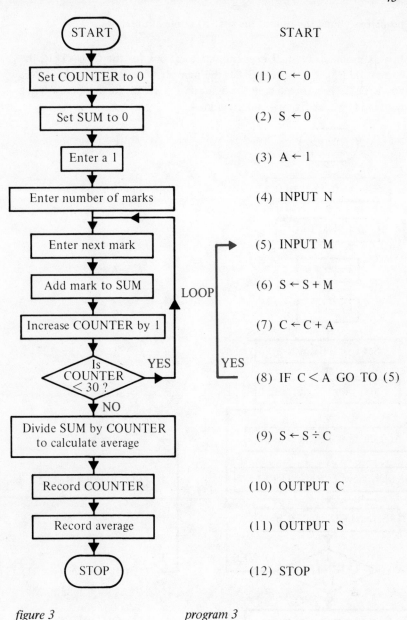

START

(1) C ← 0

(2) S ← 0

(3) A ← 1

(4) INPUT N

(5) INPUT M

(6) S ← S + M

(7) C ← C + A

LOOP

YES

(8) IF C < A GO TO (5)

(9) S ← S ÷ C

(10) OUTPUT C

(11) OUTPUT S

(12) STOP

figure 3   *program 3*

note The input device would contain the number of marks, 30, followed by the individual marks. The first input instruction (4) loads the 30 into location N, and after that the marks are loaded into location M, as they are called for by the second input instruction (5).

problem   Find the sum of the first 10 terms of the series

$$1 + \tfrac{1}{2} + \tfrac{1}{4} + \tfrac{1}{8} + \ldots$$

In this problem we shall need storage locations for the COUNTER, the current TERM, the current SUM, the number of terms required, a ONE, and a TWO. The only data in the input device will be the number of terms required (10). See figure 4 and program 4.

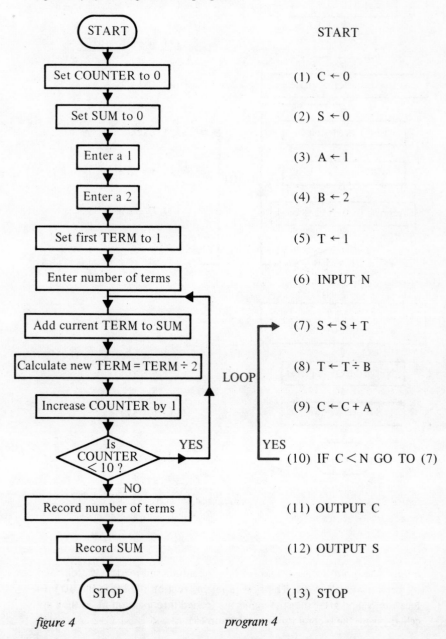

START	START
Set COUNTER to 0	(1) $C \leftarrow 0$
Set SUM to 0	(2) $S \leftarrow 0$
Enter a 1	(3) $A \leftarrow 1$
Enter a 2	(4) $B \leftarrow 2$
Set first TERM to 1	(5) $T \leftarrow 1$
Enter number of terms	(6) INPUT N
Add current TERM to SUM	(7) $S \leftarrow S + T$
Calculate new TERM = TERM ÷ 2	(8) $T \leftarrow T \div B$
Increase COUNTER by 1	(9) $C \leftarrow C + A$
Is COUNTER < 10 ?	(10) IF $C < N$ GO TO (7)
Record number of terms	(11) OUTPUT C
Record SUM	(12) OUTPUT S
STOP	(13) STOP

figure 4   *program 4*

exercise 8

1   Draw a flow chart and write a program for each of the following problems:
 (a) print out the squares of the first 20 natural numbers;
 (b) print out the reciprocals of the first 20 even numbers;
 (c) find the sum of a given list containing 40 numbers;
 (d) find the sum of the first 100 odd numbers;
 (e) find the sum of the first 10 terms of the series

$$1 + \frac{1}{1!} + \frac{1}{2!} + \frac{1}{3!} + \frac{1}{4!} + \ldots$$

In each case specify the contents of the input device. Also carry out a dry check to make sure that the program comes out of the loop at the correct time.

2   Determine precisely what each of the flow charts in figure 5 achieves.

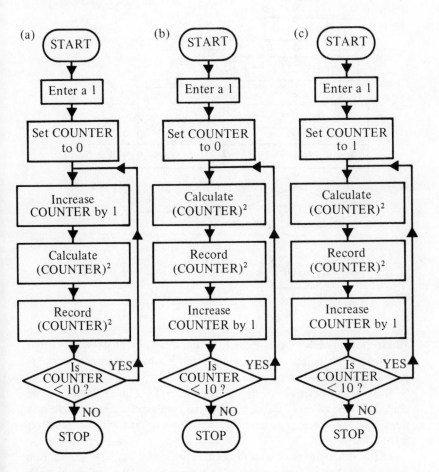

figure 5

3 Translate the flow charts in figure 5 into programs. Do a dry check on each program to confirm your results in question 2. Remember that you will need to input the number of terms (10) to a particular location, in order to ask the question, 'Is counter < 10?'

You probably discovered that the three results in question 2, although very similar, were not quite the same. Only the first flow chart correctly described how to print out the squares of the first 10 natural numbers.

The initial setting of the counter and the position of the instruction 'Increase counter by 1', are vitally important, and great care must be taken with the dry check to ensure that the exit from the loop is made at the correct time.

The flow chart in figure 6 shows an alternative version of figure 5(a).

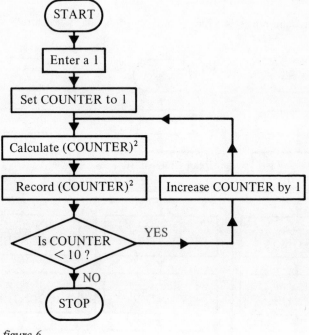

figure 6

Whilst it is usually much clearer to draw a flow chart with various instructions set to one side, a problem arises when we wish to translate such a flow chart into a program.

A program consists of a list of instructions in a linear sequence. The computer follows these through successively unless there happens to be a jump instruction which takes it back to a previous instruction or to a later instruction.

In the situation above we will need a conditional jump to take us from the decision box to the instruction 'Increase counter by 1', followed by an unconditional jump to take us from this instruction back to the instruction 'Calculate (COUNTER)^2'.

The flow chart has been re-drawn in figure 7 to show this more clearly, with the corresponding program written alongside.

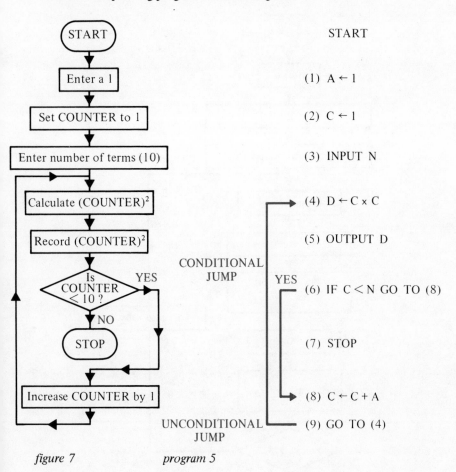

figure 7 *program 5*

note Although the STOP instruction is not necessarily the last instruction in the sequence, it will be the final instruction performed. Program 5 bypasses instruction (7) until the value of the counter has reached 10. At this stage the condition in instruction (6) will not be satisfied, and instead of jumping to instruction (8) the program will carry out the STOP, which is the next instruction.

The same ideas occur in the graph plotting situation which follows.

problem Calculate and print out the values of $1/(1 + x^2)$ for $x = -4, -3, -2, -1, 0, 1, 2,$ and 3.

Check that the flow chart in figure 8 evaluates $1/(1 + x^2)$ for all the required values of x, including −4 and +3, and then translate it into a correct program.

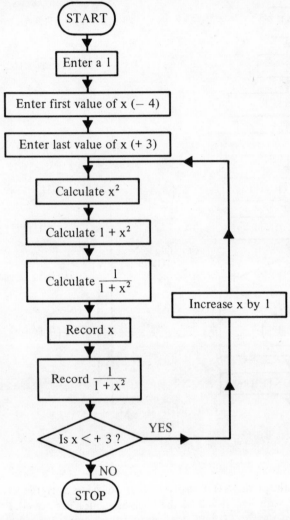

START

Enter a 1

Enter first value of x (− 4)

Enter last value of x (+ 3)

Calculate x^2

Calculate $1 + x^2$

Calculate $\dfrac{1}{1 + x^2}$

Increase x by 1

Record x

Record $\dfrac{1}{1 + x^2}$

Is x < + 3 ?   YES   NO

STOP

figure 8

First attempts at drawing this flow chart might have positioned the instruction 'Increase x by 1', either directly before the instruction 'Calculate x^2' or directly before the decision box.

In the first case, the initial calculation would have been carried out for x = −3, since the −4 value of x would have been increased by 1 before the calculation took place.

In the second case the exit from the loop would have been made when the value of x had just been increased to +3. This would have been before the calculation for that particular value of x.

It is bad practice to ignore these difficulties by adjusting the given values of x to suit *your* flow chart!

One way of getting round the difficulties in the previous situation is to use a different form of the question in the decision box.

Up till now we have tended to use questions like

'IS COUNTER LESS THAN 10?'   and   'IS $x < 3$?'

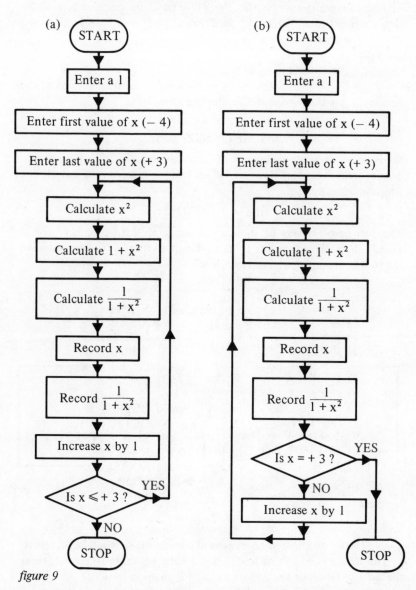

figure 9

Alternative forms of these questions may include the $\leqslant$, =, or $>$ signs, for example 'IS COUNTER = 10?' or 'IS x $\leqslant$ 3?'

Versions of the previous flow chart, using these forms are shown in figure 9.

note   By allowing the use of the $\leqslant$ sign in the first flow chart, the YES exit of the decision box is used when x = +3. This enables the calculation to be made for the final value of x.

It is interesting to note that the YES exit of the decision box, in the flow chart in figure 9(b), is only used once.

alternative conditional jump instructions

Corresponding to the use of the $\leqslant$ and = signs in the decision box in the flow chart, are the two conditional jump instructions.

$$\text{IF } C \leqslant N \text{ GO TO (5)}$$
$$\text{and } \quad \text{IF } C = N \text{ GO TO (5)}$$

Some computers allow the use of all three versions in their particular programming language. However, others may be restricted to a single one which is normally

$$\text{IF } C < N \text{ GO TO (5)}$$

The two new versions of the conditional jump instruction are used in exactly the same way as the first one. Programs 6 and 7, corresponding to the two flow charts in figure 9 use these.

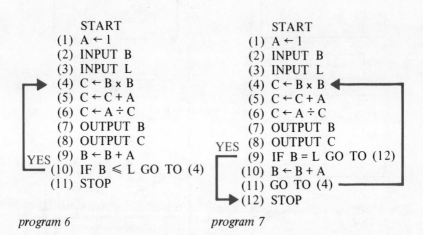

START	START
(1) A ← 1	(1) A ← 1
(2) INPUT B	(2) INPUT B
(3) INPUT L	(3) INPUT L
(4) C ← B x B	(4) C ← B x B
(5) C ← C + A	(5) C ← C + A
(6) C ← A ÷ C	(6) C ← A ÷ C
(7) OUTPUT B	(7) OUTPUT B
(8) OUTPUT C	(8) OUTPUT C
YES (9) B ← B + A	YES (9) IF B = L GO TO (12)
(10) IF B ⩽ L GO TO (4)	(10) B ← B + A
(11) STOP	(11) GO TO (4)
	(12) STOP

program 6   *program 7*

note   The input device in each case will contain the first and last values of x. The first x (−4) will be loaded into location B on instruction (2) and the last x (+3) will be loaded into location L on instruction (3).

The conditional jump instruction is of fundamental importance to almost all programming. It is therefore essential that one should be able to incorporate it correctly into a program.

One must be willing to experiment with the positioning of the decision box in a flow chart, to ensure that the exit from the loop is made at the correct time. Since every situation is likely to be different great care must be taken with the dry check.

Another situation which regularly occurs is that of raising a given number to a given power. This also incorporates all the ideas we have just met.

problem Draw a flow chart and write a program to evaluate x^n for a given $n > 0$.

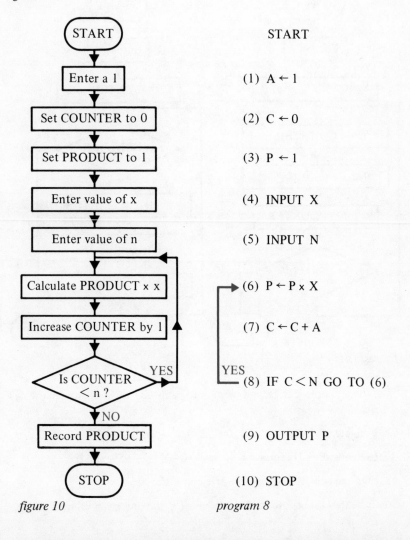

	START
Enter a 1	(1) $A \leftarrow 1$
Set COUNTER to 0	(2) $C \leftarrow 0$
Set PRODUCT to 1	(3) $P \leftarrow 1$
Enter value of x	(4) INPUT X
Enter value of n	(5) INPUT N
Calculate PRODUCT x x	(6) $P \leftarrow P \times X$
Increase COUNTER by 1	(7) $C \leftarrow C + A$
Is COUNTER < n?	(8) IF $C < N$ GO TO (6)
Record PRODUCT	(9) OUTPUT P
STOP	(10) STOP

figure 10   *program 8*

Here we can use storage locations to hold the successive PRODUCTS, and a COUNTER to record the number of times the PRODUCT has been multiplied by x. We shall also need a question such as 'Is COUNTER < n?' to determine whether or not the required power has been reached.

If the PRODUCT location initially contains ONE, then the COUNTER must be set to zero to keep in step with the power. This is shown in the flow chart in figure 10 and program 8.

exercise 9

1   Determine what each of the flow charts in figure 11 achieves and translate each into a corresponding program.

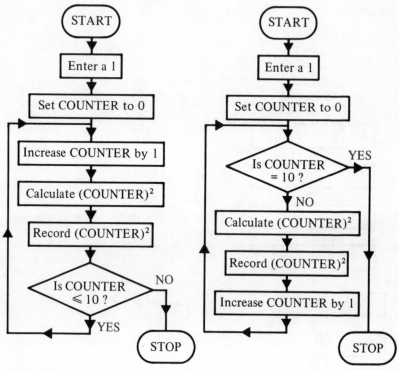

figure 11

Draw flow charts and write programs for each of the following situations.

2   Evaluate x^3 for $x = -2, -1, 0, 1, 2, 3, 4, 5$, and 6.

3   Evaluate $x(x - 1)$ for $x = 0.40, 0.41, 0.42, \ldots, 0.59$, and 0.60.

4   Find the sum of $1 + 2 + 2^2 + 2^3 + \ldots + 2^9$

5   Evaluate $(1 + x)^n$ for a fixed value of x and any given value of $n > 0$.

6　Evaluate $P(1 + R/100)^{10}$ for given values of P and R.

7　Evaluate $a(r^n - 1)/(r - 1)$, for fixed values of a, and r, and a given value of $n > 0$.

8　A man invests £10 each month in an insurance scheme. At the end of each month interest is added on to the existing capital at the rate of 5% per year. Calculate the total amount that will be invested at the end of a 12 month period.

9　Find the average of the first 10 cubes.

10　Find the largest of 10 given different numbers.

false data

Earlier in this book when we had a list of data to enter into the computer, we allowed a question such as 'Is there any more data?', to determine whether we had reached the end of the list. One of the snags with this question is that the computer will not be able to distinguish between a piece of data which happens to be zero, and the fact that there is no more data to be entered.

If the data input were on paper tape, which had the six numbers 3, 15, 0, 0, 0, 17, punched on it in sequence, it would look like figure 12.

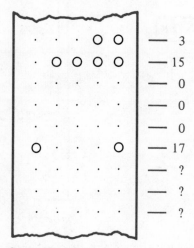

figure 12

It would be impossible to decide, after entering the 17, whether the line which followed with no holes in it, was in fact a piece of data, zero, or whether it just happened to be the continuation of the tape. A similar situation would exist where the input was punched cards.

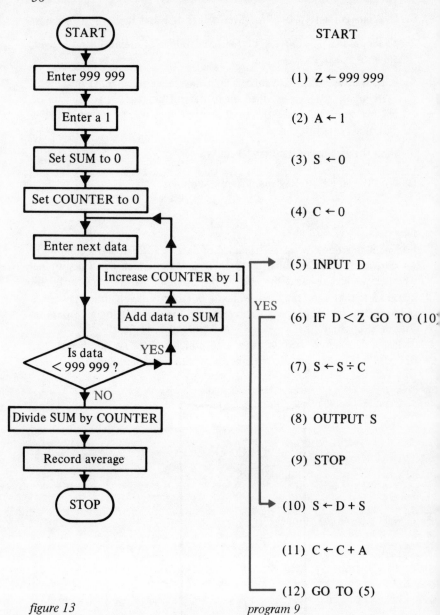

START

(1) Z ← 999 999

(2) A ← 1

(3) S ← 0

(4) C ← 0

(5) INPUT D

(6) IF D < Z GO TO (10)

(7) S ← S ÷ C

(8) OUTPUT S

(9) STOP

(10) S ← D + S

(11) C ← C + A

(12) GO TO (5)

figure 13   *program 9*

There are various ways round this problem, and one we have already met, where we know how many items there were in the original list. In that case we used a COUNTER, which increased by 1, each time an item was entered. By asking a question such as

Is COUNTER < number of items?

we were able to decide whether there was any more data.

Unfortunately it is not always possible to know how much data there is, and often it would be very time consuming to count it all manually in order to find out. In this type of situation it is normal practice to add a piece of meaningless data at the end of the list.

For example if we wanted to find the average of a list of marks, which we happened to know to be positive, we could add a negative number such as −1 as the final mark. The question

Is data < 0?

would only have the answer YES when all the other data had been entered.

If the data contained both positive and negative numbers we might add a number such as 1 000 000 and ask the question

Is data < 999 999?

Assuming none of the data was larger than this number, the answer to the question would be YES, until we reached the final item.

The idea of using false data to end a loop is shown in the flow chart in figure 13 and program 9 for the next problem.

problem Calculate the average of a given list of numbers.

If we assume that all the numbers are less than 999 999 we may add 1 000 000 as the final piece of data. The list together with this number would be held in the input device.

division

Another situation which necessitates the idea of a conditional jump instruction is that of division. Whilst we have allowed 'A ← B ÷ C' as an instruction that the computer can carry out, at some stage it will have been broken down into a series of steps based on addition. We saw how this was done for multiplication on page 44.

Division can be thought of either as a process based on repeated subtraction, or as one based on repeated addition. If we wished to calculate the QUOTIENT and REMAINDER for 143 ÷ 12 we could either count the number of times we were able to subtract 12 from 143, or we could find the largest number of 12's which when added together were not greater than 143.

In the first case the remainder is the number we are left with when we have done the last subtraction. In the second case we have to subtract the result from 143. The two methods are described in the flow charts in figures 14 and 15.

note It is necessary to have the DECISION BOX before the main calculation to cater for the case where the numerator was originally less than the denominator. For example in 11 ÷ 43 the QUOTIENT would be

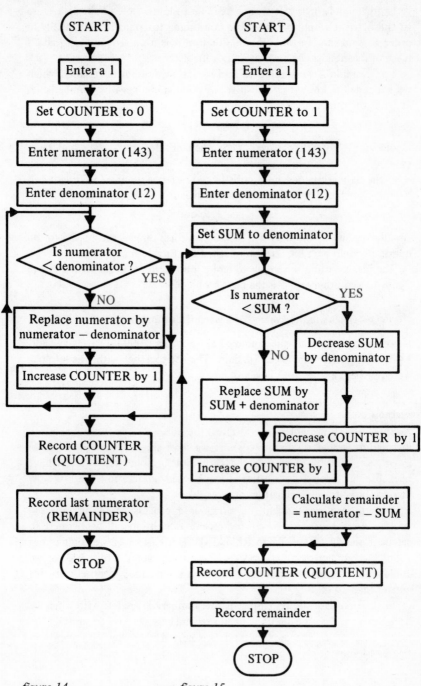

figure 14

figure 15

0 and the REMAINDER 11. In the flow chart in figure 15 it is easier to wait until the SUM becomes larger than the NUMERATOR, than to have to ask the question, 'Is the difference between the numerator and the sum less than the denominator?'.

The programs for each method are 10 and 11. In program 11 we will allow the instruction 'S ← D' for 'set SUM to denominator'.

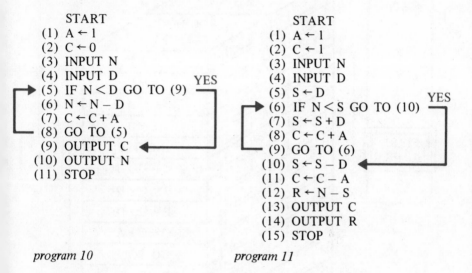

```
        START                          START
    (1) A ← 1                      (1) A ← 1
    (2) C ← 0                      (2) C ← 1
    (3) INPUT N                    (3) INPUT N
    (4) INPUT D           YES      (4) INPUT D                    YES
  ┌▸(5) IF N < D GO TO (9) ──┐     (5) S ← D
  │ (6) N ← N − D            │   ┌▸(6) IF N < S GO TO (10) ──┐
  │ (7) C ← C + A            │   │ (7) S ← S + D             │
  └─(8) GO TO (5)            │   │ (8) C ← C + A             │
    (9) OUTPUT C  ◂──────────┘   └─(9) GO TO (6)             │
   (10) OUTPUT N                    (10) S ← S − D  ◂─────────┘
   (11) STOP                        (11) C ← C − A
                                    (12) R ← N − S
                                    (13) OUTPUT C
                                    (14) OUTPUT R
                                    (15) STOP

      program 10                       program 11
```

We shall leave the problem of carrying out a division correct to a given number of decimal places until chapter 7. In the next chapter the idea of allowing × and ÷ as permissible instructions will be extended to include instructions such as

A ← √B meaning replace the contents of location A by the square root of the contents of location B

or P ← sin Q meaning replace the contents of P by the sine of the contents of Q

or N ← log R meaning replace the contents of N by the logarithm of the contents of R

As the computer has developed it has been possible to allow the use of more and more sophisticated instructions within the program. These are then translated, by the COMPILER, into the actual steps necessary for the particular calculation.

project the matchbox computer − false data
Figure 16 shows a list of data, prepared on paper tape, waiting to be called for by the input device. The data consists of a set of positive numbers, and the false data −1 has been added to indicate the end of the list.

Load the given program into the program store of your matchbox computer and place a set of numbers into your input device. Make sure you have −1 as the final piece of data. Carry out the program and find out what calculation is being performed on your set of numbers.

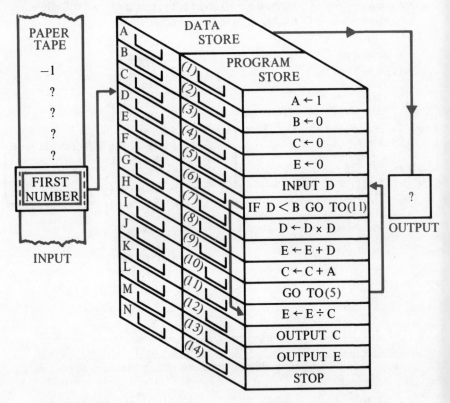

figure 16

Choose a problem you would like to program. Write out the list of instructions and load them into your program store. Decide on the data necessary, and load this, in order, into the input device. Give your computer to a friend. Tell him where to find the first instruction, and ask him to carry out your program. Did he obtain the required results?

6 subroutines for functions

We have already seen that whilst we may allow instructions such as 'A ← B × C' and 'P ← S ÷ T', the computer in fact has to carry out a small set of instructions to achieve the desired result. At some stage within the computer, each calculation will have been broken down into programs based on addition.

In a similar way, provided we can express such things as $\sqrt{B}$, or sin C, as calculations based on addition, we may allow instructions like 'A ← $\sqrt{B}$' and 'D ← sin C'.

The sets of instructions to carry out these expressions are known as SUBROUTINES. These would form a normal part of the SOFTWARE of the computer. In other words having written and perfected them for his particular machine, the manufacturer would provide them, when he sold the computer, to enable the user to write shorter and more meaningful programs, incorporating this type of instruction.

Let us first look in detail to see whether it is possible to calculate $\sqrt{B}$ in this way, and then see how the instruction 'A ← $\sqrt{B}$' is actually carried out by the computer.

square root

Many mathematical expressions have been calculated and can be obtained by looking up the relevant information in a set of tables. One might ask why the computer cannot store the information from a set of four-figure tables, as individual entries. In fact it would need considerably more storage to do this, than to store the list of instructions to calculate the particular expression. Although it would take rather more time to do the calculation, than to look up the information, the cost of storage in a computer is vastly more expensive than the cost of running the machine.

problem   Find $\sqrt{12}$.

One of the delightful facilities of a computer is its ability to produce results to any required degree of accuracy. Often a scientist will require his results correct to many significant figures.

We can use the computer's ability to repeat a given set of instructions and to ask particular questions to achieve the desired accuracy.

The method set out below to calculate $\sqrt{12}$ uses these ideas and was devised by a man called HERO nearly 2000 years ago.

A first estimate for $\sqrt{12}$ might be 3, since we know $3 \times 3 = 9$, and $4 \times 4 = 16$. A rectangle whose area is 12 square units, with one side of 3 units, would have the other side equal to 4 units.

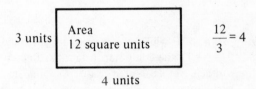

$$3 \text{ units} \quad \boxed{\begin{array}{l} \text{Area} \\ 12 \text{ square units} \end{array}} \qquad \frac{12}{3} = 4$$

4 units

Since in finding $\sqrt{12}$, we are trying to find the side of a square whose area is 12 square units, a better estimate than 3 would be the average of 3 and 4, namely 3·5 units. Using this as one side of the rectangle, we can calculate the other side

$$\frac{12}{3 \cdot 5} \simeq 3 \cdot 4 \qquad 3 \cdot 5 \text{ units} \quad \boxed{\begin{array}{l} \text{Area} \\ 12 \text{ square units} \end{array}}$$

3·4 units

We can continue this process, calculating each new estimate from the previous one.

Estimate 3·45 $\qquad \dfrac{12}{3 \cdot 45} \simeq 3 \cdot 48 \qquad$ Average $\frac{1}{2}(3 \cdot 45 + 3 \cdot 48) = 3 \cdot 465$

In general terms, if x were the estimate, then a better estimate would be

$$\frac{1}{2}\left(x + \frac{N}{x}\right)$$

The required result would be obtained when two successive estimates were sufficiently close for the required degree of accuracy.

The flow chart in figure 1 summarizes this line of attack.

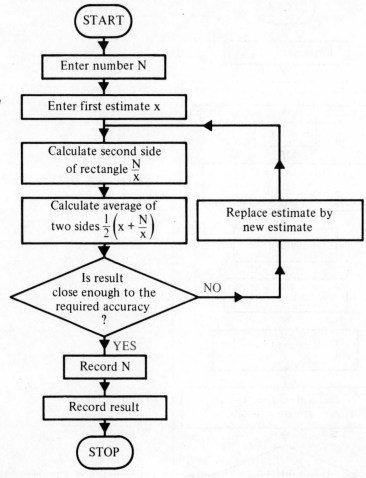

figure 1

degree of accuracy

In asking whether the result is to the required accuracy, we are really asking whether the result is correct to a certain number of decimal places. We can do this by using either of the following questions.

Is the difference between two successive estimates $< 0 \cdot 00001$?

or   Is the difference between the $(\text{estimate})^2$ and the number $< 0 \cdot 00001$? For either we would need to modify the flow chart above by having instructions to enter the required degree of accuracy ($0 \cdot 00001$) and to calculate the particular difference.

The modified flow chart is shown in figure 2 and program 1 is the corresponding program.

64

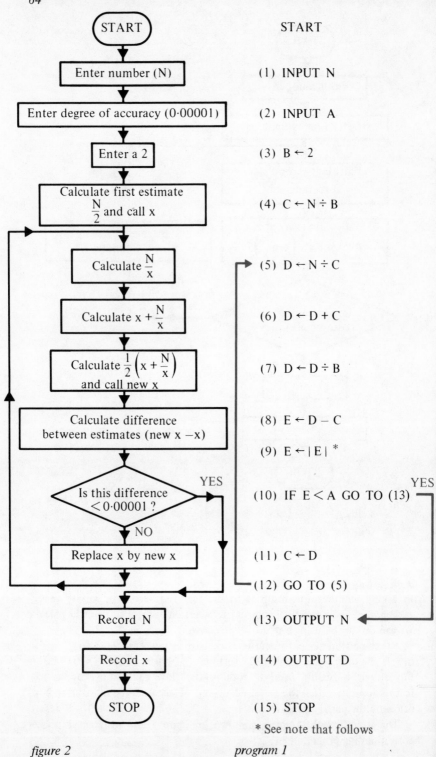

figure 2

START

(1) INPUT N

(2) INPUT A

(3) B ← 2

(4) C ← N ÷ B

(5) D ← N ÷ C

(6) D ← D + C

(7) D ← D ÷ B

(8) E ← D − C

(9) E ← | E | *

(10) IF E < A GO TO (13)

(11) C ← D

(12) GO TO (5)

(13) OUTPUT N

(14) OUTPUT D

(15) STOP

* See note that follows

program 1

note A TWO is needed in order to find the average and it is also used to calculate the first estimate. The computer has to be told what the first estimate is, and although we use half the original number the choice is purely arbitrary. Any reasonable number, except 0 could have been used instead.

A problem exists in the question 'Is the difference < 0.00001?' since the word 'difference' usually implies that we take the smaller number from the larger one. The computer will only do the subtraction that it is given, unless it is first programmed to find which of the two numbers is the larger. Alternatively the computer can find the MODULUS, or positive value, of the difference, as is done in program 1. Flow charts showing how each of these methods are carried out are shown in figure 3. A third way

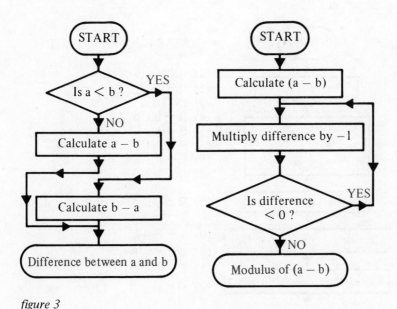

figure 3

of dealing with the problem is by using the conditional jump instruction,

$$\text{IF } |a - b| < 0.00001 \text{ GO TO (13)}$$

Some machines allow this facility of asking questions about a modulus, whilst others allow the direct calculation of a modulus. If this is not the case, then a few instructions in the program, corresponding to either of the flow charts in figure 3, would be necessary.

Once it is appreciated how $\sqrt{n}$ and $|x|$ can be calculated we may allow 'A $\leftarrow \sqrt{B}$' and 'P $\leftarrow |Q|$' as permissible instructions. The following situation uses the first of these instructions.

problem   Draw a flow chart and write a program to evaluate

$$\frac{-b \pm \sqrt{(b^2 - 4ac)}}{2a}$$

(See figure 4 and program 2.) Whilst we have allowed the instruction '$D \leftarrow \sqrt{D}$' in (9), the computer would actually carry out a square root subroutine, which would be very similar to the program on page 64.

On meeting instruction (9) it would 'GO TO $\sqrt{}$ SUBROUTINE' carry out that particular set of instructions and then return to instruction (10). The various subroutines would be stored in another part of the program store, and in this case the last instruction of the subroutine would have to be 'GO TO (10)'.

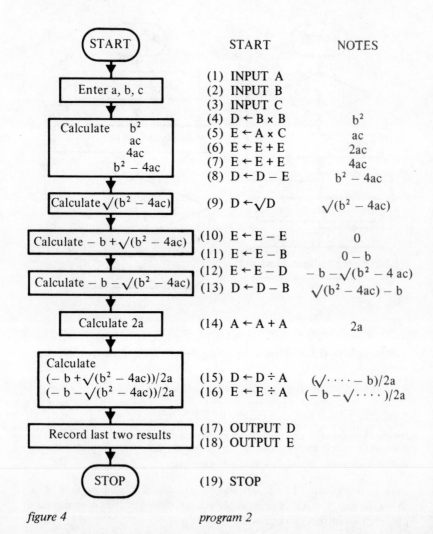

START	START	NOTES
Enter a, b, c	(1) INPUT A	
	(2) INPUT B	
	(3) INPUT C	
Calculate b^2	(4) $D \leftarrow B \times B$	b^2
ac	(5) $E \leftarrow A \times C$	ac
4ac	(6) $E \leftarrow E + E$	2ac
$b^2 - 4ac$	(7) $E \leftarrow E + E$	4ac
	(8) $D \leftarrow D - E$	$b^2 - 4ac$
Calculate $\sqrt{(b^2 - 4ac)}$	(9) $D \leftarrow \sqrt{D}$	$\sqrt{(b^2 - 4ac)}$
Calculate $-b + \sqrt{(b^2 - 4ac)}$	(10) $E \leftarrow E - E$	0
	(11) $E \leftarrow E - B$	$0 - b$
	(12) $E \leftarrow E - D$	$-b - \sqrt{(b^2 - 4ac)}$
Calculate $-b - \sqrt{(b^2 - 4ac)}$	(13) $D \leftarrow D - B$	$\sqrt{(b^2 - 4ac)} - b$
Calculate 2a	(14) $A \leftarrow A + A$	2a
Calculate $(-b + \sqrt{(b^2 - 4ac)})/2a$ $(-b - \sqrt{(b^2 - 4ac)})/2a$	(15) $D \leftarrow D \div A$	$(\sqrt{\cdots} - b)/2a$
	(16) $E \leftarrow E \div A$	$(-b - \sqrt{\cdots})/2a$
Record last two results	(17) OUTPUT D	
	(18) OUTPUT E	
STOP	(19) STOP	

figure 4   *program 2*

We can simulate this on the matchbox computer by adding an additional storage unit to hold the subroutines.

project   the matchbox computer – subroutines

In figure 5 the program store of the matchbox computer has been split into two parts. One contains a subroutine, and the other holds the main program.

What calculation does the subroutine perform on the contents of location X? What is the effect of the main program when the subroutine is used?

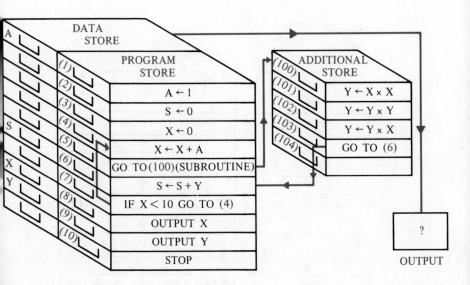

figure 5

If the subroutine in locations 100–104 had been

$$
\begin{array}{ll}
(100) & Y \leftarrow A \div X \\
(101) & Y \leftarrow Y \times Y \\
(102) & Y \leftarrow Y \div X \\
(103) & Y \leftarrow Y \times Y \\
(104) & \text{GO TO (6)}
\end{array}
$$

what would have been the effect of the program?

The main program in locations 1-10 will calculate the sum of the first ten values of any expression whose subroutine is held in locations 100 onwards.

trigonometrical functions

Many mathematical functions can be expressed as infinite power series. By taking the sum of a sufficiently large number of terms in the series, it is possible to evaluate such functions as sin x, or log (1 + x) to any required degree of accuracy.

The series for sin x is

$$x - \frac{x^3}{3!} + \frac{x^5}{5!} - \frac{x^7}{7!} + \frac{x^9}{9!} \ldots$$

This is convergent for all values of x, providing x is measured in radians.

problem   Evaluate sin x, correct to four decimal places, from the above series.

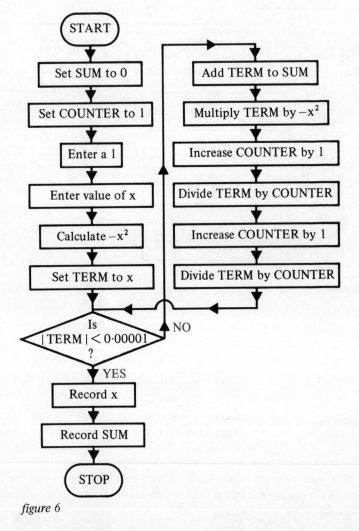

figure 6

In this situation we can obtain each term from the previous term, and we can terminate the summation when the modulus of any given term is less than 0·00001. This is described in the flow chart in figure 6. Program 3 is for evaluating sin x. The dry check indicates the contents of the storage location in the particular instruction, on successive loops.

START	RESULTS OF SUCCESSIVE LOOPS				
(1) $S \leftarrow 0$	0				
(2) $C \leftarrow 1$	1				
(3) $A \leftarrow 1$	1				
(4) $E \leftarrow 0.00001$	0·00001				
(5) INPUT X	x				
(6) $B \leftarrow X \times X$	x^2				
(7) $B \leftarrow S - B$	$0 - x^2$				
(8) $T \leftarrow X$ YES	x				
(9) IF $|T| < E$ GO TO (17)	NO	NO	NO	...	YES
(10) $S \leftarrow S + T$	x	$x - \dfrac{x^3}{3!}$	$x - \dfrac{x^3}{3!} + \dfrac{x^5}{5!}$		
(11) $T \leftarrow T \times B$	$-x^3$	$+\dfrac{x^5}{3!}$	$-\dfrac{x^7}{5!}$		
(12) $C \leftarrow C + A$	2	4	6		
(13) $T \leftarrow T \div C$	$-\dfrac{x^3}{2}$	$\dfrac{x^5}{4!}$	$-\dfrac{x^7}{6!}$		
(14) $C \leftarrow C + A$	3	5	7		
(15) $T \leftarrow T \div C$	$-\dfrac{x^3}{2.3}$	$\dfrac{x^5}{5!}$	$-\dfrac{x^7}{7!}$	...	etc
(16) GO TO (9)					
(17) OUTPUT X				x	
(18) OUTPUT S				$x - \dfrac{x^3}{3!} + \dfrac{x^5}{5!} - \dfrac{x^7}{7!} \cdots$	
(19) STOP					

program 3

exercise 10

Write a program to calculate each of the given functions from their series.

1   $\cos x = 1 - \dfrac{x^2}{2!} + \dfrac{x^4}{4!} - \dfrac{x^6}{6!} + \dfrac{x^8}{8!} - \ldots$

2   $e^x = 1 + x + \dfrac{x^2}{2!} + \dfrac{x^3}{3!} + \dfrac{x^4}{4!} + \ldots$

3   $\log(1 + x) = x - \dfrac{x^2}{2} + \dfrac{x^3}{3} - \dfrac{x^4}{4} + \dfrac{x^5}{5} - \ldots$ where $-1 < x \leqslant 1$

4   $(1 + x)^n = 1 + nx + \dfrac{n(n-1)}{2!} x^2 + \dfrac{n(n-1)(n-2)}{3!} x^3 + \ldots$ where $-1 < x < 1$

You may use an instruction such as (9) above in order to determine the required degree of accuracy.

In actual practice the series given in the last exercise may converge very slowly. As a result more complicated variations of these are generally used, but the principle is the same. The individual manufacturers will have perfected the various subroutines for these expressions, and they will be used to carry out an instruction such as 'P ← sin Q'.

In calculating 'a sin C − b' program 4 might be used.

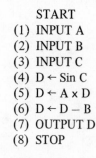

START
(1) INPUT A
(2) INPUT B
(3) INPUT C
(4) D ← Sin C
(5) D ← A x D
(6) D ← D − B
(7) OUTPUT D
(8) STOP

program 4

If we were to use a subroutine for evaluating the sine function then program 5 would be the revised program. The subroutine would be written

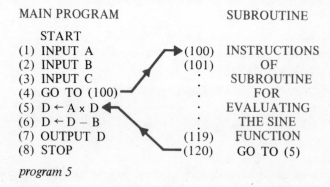

MAIN PROGRAM   SUBROUTINE

START
(1) INPUT A   →(100) INSTRUCTIONS
(2) INPUT B   (101)   OF
(3) INPUT C   ·   SUBROUTINE
(4) GO TO (100)   ·   FOR
(5) D ← A x D   ·   EVALUATING
(6) D ← D − B   ·   THE SINE
(7) OUTPUT D   (119) FUNCTION
(8) STOP   (120) GO TO (5)

program 5

to evaluate the sine of the angle held in location C and to place the result in location D. In addition an unconditional jump instruction, returning the subroutine to the main program, would have to be added as the last instruction of the subroutine.

A problem will arise if we want to use the subroutine at two different places in the main program. How can we arrange for the subroutine to know at which point to re-enter the main program?

project   the matchbox computer − arithmetic unit

The calculations in the situations that follow involve the evaluation of various trigonometric expressions and also the use of logarithms. Designate an area of your matchbox computer to be the ARITHMETIC UNIT and in addition to giving it the ability to do +, −, x, and ÷, provide it with a set of three- or four-figure tables, or a slide rule. This will allow you to evaluate most reasonable expressions, without having to bother with a comprehensive set of subroutines.

What calculation is being carried out by the program in figure 7?

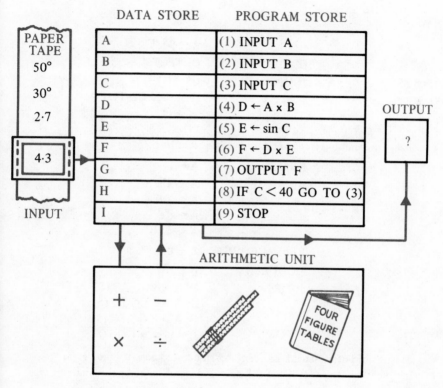

figure 7

exercise 11

Write programs to calculate the following:

1 The area of triangle, given two sides and the included angle.

2 The hypotenuse of a right-angled triangle, given the other two sides.

3 The side, a, of a triangle from the formula $a^2 = b^2 + c^2 - 2bc \cos A$.

4 The side, b, of a triangle from the formula $a/\sin A = b/\sin B$.

5 The area of a triangle from the formula $\Delta = \sqrt{[s(s-a)(s-b)(s-c)]}$ where $s = \frac{1}{2}(a + b + c)$.

full summary of allowable instructions

set of 8 basic instructions

INPUT	INPUT A
ARITHMETIC	$A \leftarrow B + C$
	$A \leftarrow B - C$
	$A \leftarrow B \times C$
	$A \leftarrow B \div C$
OUTPUT	OUTPUT A
JUMPS	GO TO (3)
	IF $A < B$ GO TO (5)

other instructions depending on available machine

INPUT	$A \leftarrow 1$
ARITHMETIC	$A \leftarrow A \times A$
	$A \leftarrow B + A$
	$A \leftarrow \sqrt{B}$
	$A \leftarrow \sin B$
REPLACEMENT	$A \leftarrow B$
MODULUS	$A \leftarrow |A|$
JUMPS	IF $A = B$ GO TO (5)
	IF $A > B$ GO TO (5)
	IF $A \geq B$ GO TO (5)
	IF $|A| < |B|$ GO TO (5) etc.

Most computer languages can be used to accept the eight basic instructions. The use of the others will depend on the particular language and machine, and on whether suitable subroutines are available.

7 from problem to program

This chapter will be devoted to a set of worked problems. In each case the problem will be discussed, and the individual steps looked at in detail. Often a small part of the flow chart, or a few instructions in the program, will be developed to illustrate an important idea.

It is hoped in this way to show how such problems can be solved with the aid of a computer, and also to see the type of thinking necessary to construct the appropriate program.

worked problem 1   Arrange three given numbers in ascending order.

The amount of arithmetic in this problem is negligible, but we can use the computer's ability to ask questions and to compare two numbers.

One method of tackling this problem, which can also be extended to several numbers, is to take any number and compare it with each of the others in turn. The comparison is to decide which of the two numbers is the smaller. If the first number is larger than, or equal to the second, then the two are interchanged. If it is smaller, then it is compared with the next number on the list. This is shown in the flow chart and program below.

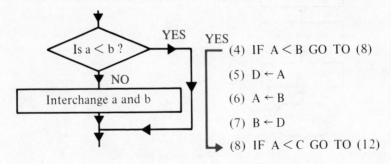

This process continues until the first number or its replacement has been compared with all the numbers in the list. The number we finish up with will be the smallest of the set. This can be repeated to find the next smallest and so on.

For each successive number there will be one fewer comparison than for the previous number. The program will contain several very similar lists of instructions corresponding to each new comparison. The full flow chart and program are shown in figure 1 and program 1.

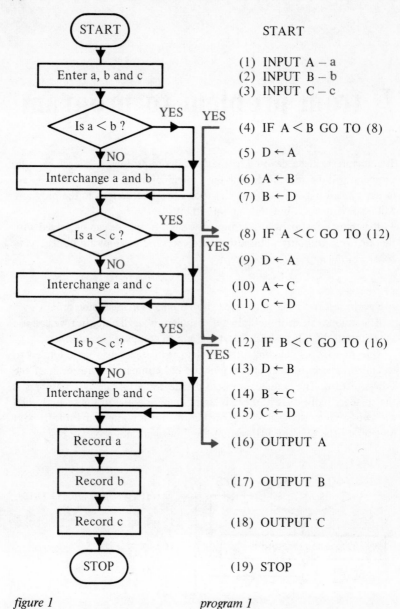

figure 1   *program 1*

worked problem 2 Calculate x^n for any value of n.

In this problem, as in the previous one, the main difficulty is to remember all the possibilities that might occur. Here n could be positive, negative or zero.

We saw on page 53 how to write a program to calculate x^n when n was positive. We can modify this by making a decision, before the calculation, asking if n is negative. This will then allow for the two main possibilities. By re-positioning the original decision, as shown in figure 2 we can also cater for the case where n actually equals zero.

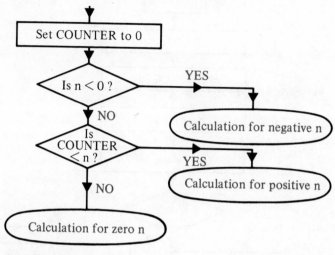

figure 2

If n is negative we can think of the situation as raising $\frac{1}{x}$ to the particular power, or continually dividing 1 by x until we reach the required value. We can determine this point by reducing the COUNTER each time by one, until its value is n. This is shown in the flow chart with its corresponding program.

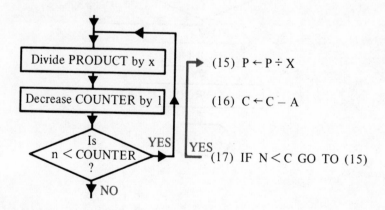

The full flow chart is shown in figure 3. Since a program is a linear sequence of instructions, we shall need to add the parts of the program corresponding to the left and right hand parts of the flow chart, after that corresponding to the central part of the flow chart. This will necessitate

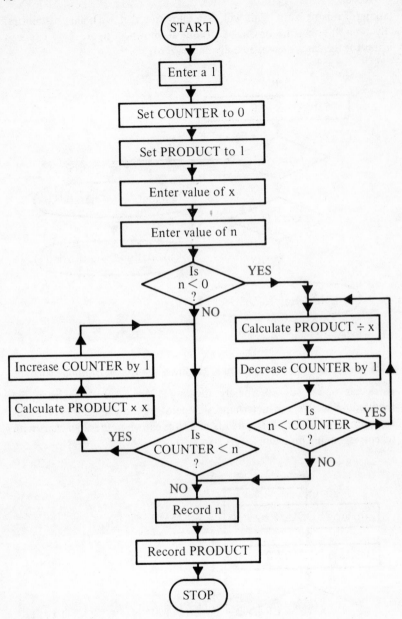

figure 3

the use of two unconditional jump instructions to return to the relevant earlier part of the program—see program 2 with appropriate notes. Carry out a DRY CHECK on this program giving x the value 2, and n the values +5, 0, and −3 in turn. In the second case you should use instructions (1) to (10) in sequence without needing instructions (11) to (17)

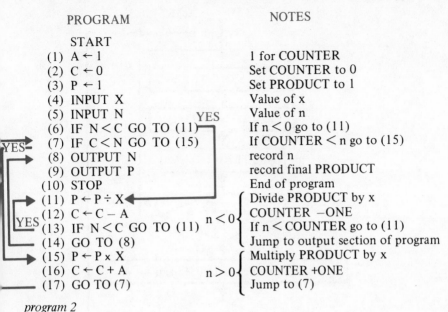

PROGRAM	NOTES
START	
(1) A ← 1	1 for COUNTER
(2) C ← 0	Set COUNTER to 0
(3) P ← 1	Set PRODUCT to 1
(4) INPUT X	Value of x
(5) INPUT N	Value of n
(6) IF N < C GO TO (11)	If n < 0 go to (11)
(7) IF C < N GO TO (15)	If COUNTER < n go to (15)
(8) OUTPUT N	record n
(9) OUTPUT P	record final PRODUCT
(10) STOP	End of program
(11) P ← P ÷ X	Divide PRODUCT by x
(12) C ← C − A	COUNTER −ONE
(13) IF N < C GO TO (11)	If n < COUNTER go to (11)
(14) GO TO (8)	Jump to output section of program
(15) P ← P x X	Multiply PRODUCT by x
(16) C ← C + A	COUNTER +ONE
(17) GO TO (7)	Jump to (7)

program 2

worked problem 3   Print out the sum of the first 10 terms of the series
$1 + \frac{1}{2} + \frac{1}{3} + \frac{1}{4} \ldots$
Extend your program to print out the sum of the first 20 terms, 30 terms,
. . . up to 100 terms.

A similar problem to the first part was set out on page 37. A
COUNTER was used to show how many terms had been added to the sum
at any stage. In this problem we will need two COUNTERS, the first to
count up to ten for each print out and the second to record the number of
print outs.

The heart of the problem is shown in the part of the flow chart in
figure 4.

Since each new term corresponds to its position, for example the fifth
term is $\frac{1}{5}$, the simplest way of asking the second question will be

Is the number of terms < 100?

This will mean that the TERM COUNTER will go from 1 to 100. In
addition we will require a LOOP COUNTER which goes from 1 to 10 to
determine the exit from the first loop.

After each print out of the appropriate sum and before starting the first
loop again, we will need to reset this second counter to zero.

An alternative way round the difficulty of the two counters, is to
modify the first question after each print out. It could read successively,
IS COUNTER < 10?, IS COUNTER < 20?, IS COUNTER < 30? etc. This
would need a store holding first 10, then 20, 30, 40, . . ., to act as an
INDICATOR.

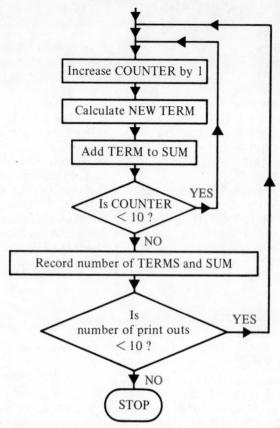

figure 4

In either method great care must be taken to give the COUNTER, TERM, SUM and PRODUCT the correct initial values. We must also remember to enter the items of data necessary for the calculation, 1, 10 and 100.

The flow charts in figure 5 describe each of the methods outlined on page 77. Several problems use the idea of a loop within a loop. Note how in the first flow chart it is possible to use the initial setting of the LOOP COUNTER to reset its value after each print out. By placing the instruction between the two returns from the LOOPS, it is only used by the outer one.

This can also be done in the second flow chart by altering the initial INDICATOR setting to 0 and moving the 'Increase INDICATOR by 10' instruction to a position between the two returns from the LOOPS.

Programs 3 and 4 correspond to the two methods. Run one program and find out the sum of the series for a very large number of terms. Can you say that the sum will tend to some particular number as the number of terms becomes infinite?

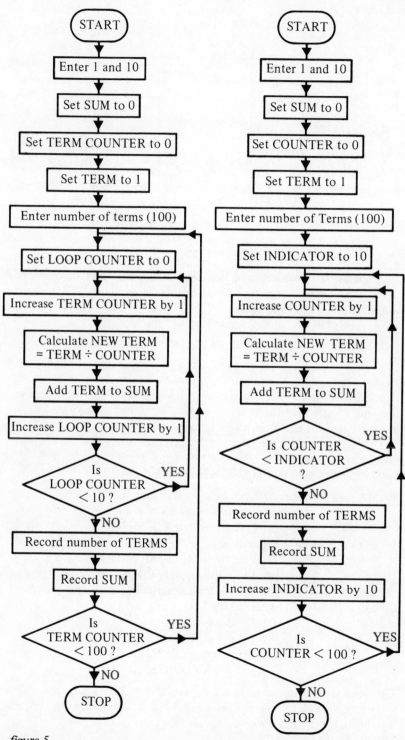

figure 5

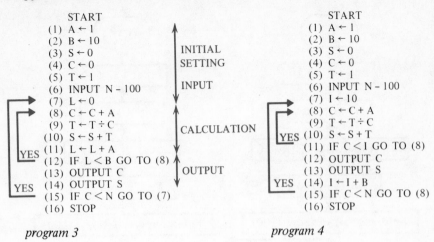

program 3   program 4

A commercial situation using the idea of a loop within a loop occurs in regular investment situations or in mortgage repayments. The problem which follows is typical of these.

worked problem 4   A man decides to invest a regular sum of money each month over a period of 10 years. Interest is added to the existing capital at the end of each month. It is required to know how much savings the investor has at the end of each year in the 10-year period.

Let us assume the interest rate is 5% per annum and that the man can save £8 per month.

The calculation in this problem is outlined in figure 6.

We will need to know   (i) The regular monthly payment P – £8
                                  (ii) The number of years in the term N – 10 years
                                  (iii) The interest rate R – 5%

We will require three fixed constants
                                  (i) 1 for use with the COUNTERS
                                  (ii) 12 for the monthly COUNTER
                                  (iii) 100 for the calculation of the monthly interest

Storage locations will be needed for
                                  (i) The current sum S
                                  (ii) The calculated interest I
                                  (iii) The monthly COUNTER C
                                  (iv) The yearly COUNTER D

The initial settings for the SUM and the two COUNTERS will be zero.

Having decided on the precise requirement a detailed flow chart can now be drawn (see figure 7 and program 5).

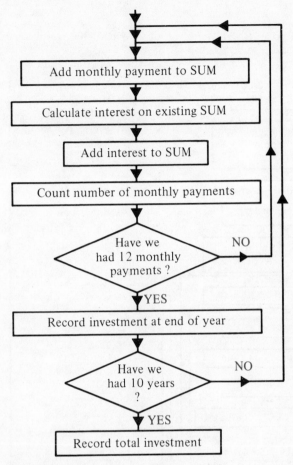

Add monthly payment to SUM

Calculate interest on existing SUM

Add interest to SUM

Count number of monthly payments

Have we had 12 monthly payments ?

NO

YES

Record investment at end of year

Have we had 10 years ?

NO

YES

Record total investment

figure 6

worked problem 5 Calculate the MEAN and STANDARD DEVIATION of a given list of numbers.

Let us assume that all numbers in the given list are less than 10 000. Since we do not know how many there are, we will use the false data 1 000 000, added at the end of the list and the question

Is data < 999 999?'

to determine when all the data has been entered.

The MEAN of the set of numbers is their total SUM divided by the number of numbers. We will need a counter to record each time a new number is entered. The STANDARD DEVIATION of the set of numbers is given by the formula,

$$\sqrt{\left(\frac{\Sigma x^2}{N} - \left(\frac{\Sigma x}{N}\right)^2\right)}$$

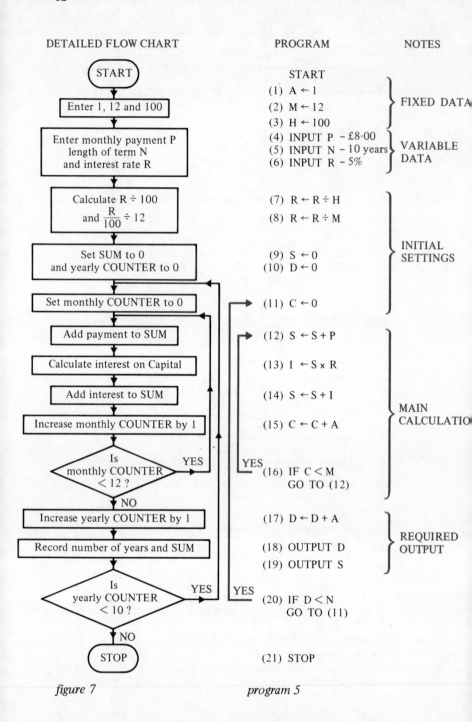

DETAILED FLOW CHART	PROGRAM	NOTES

START

START
(1) A ← 1
(2) M ← 12
(3) H ← 100 } FIXED DATA

Enter 1, 12 and 100

Enter monthly payment P
length of term N
and interest rate R

(4) INPUT P – £8·00
(5) INPUT N – 10 years
(6) INPUT R – 5% } VARIABLE DATA

Calculate R ÷ 100
and $\frac{R}{100}$ ÷ 12

(7) R ← R ÷ H
(8) R ← R ÷ M

Set SUM to 0
and yearly COUNTER to 0

(9) S ← 0
(10) D ← 0 } INITIAL SETTINGS

Set monthly COUNTER to 0

(11) C ← 0

Add payment to SUM

(12) S ← S + P

Calculate interest on Capital

(13) I ← S × R

Add interest to SUM

(14) S ← S + I

Increase monthly COUNTER by 1

(15) C ← C + A

Is
monthly COUNTER
< 12 ?   YES

YES
(16) IF C < M
GO TO (12) } MAIN CALCULATION

NO

Increase yearly COUNTER by 1

(17) D ← D + A

Record number of years and SUM

(18) OUTPUT D
(19) OUTPUT S } REQUIRED OUTPUT

Is
yearly COUNTER
< 10 ?   YES

YES
(20) IF D < N
GO TO (11)

NO

STOP

(21) STOP

figure 7   *program 5*

where Σx^2 is the sum of the squares of each of the numbers, Σx is their SUM, and N is the number of numbers.

In this situation the main problem is the actual calculation. One method is outlined in figure 8.

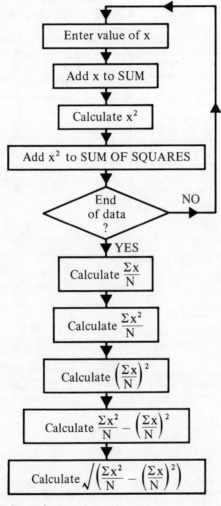

figure 8

The full flow chart falls into two parts (see figure 9). The first is where the data is entered and counted, and where the two sums are totalled. The second is where the main calculation is made.

note   It is necessary to ask the question about the data immediately it has been entered. We do not want to add the false data to the sums, or count it as one of the items.

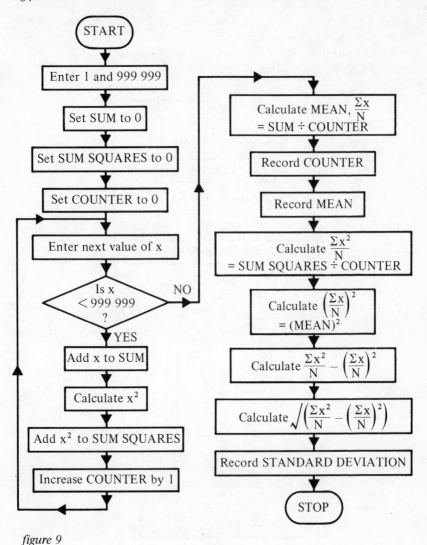

figure 9

The program will require an unconditional jump instruction, after 'Increase COUNTER by 1' to return to the 'Enter next value of x' instruction. Program 6 is the full program, with notes shown alongside.

worked problem 6   Calculate the gradients of the chords of the curve $y = x^2$ for the intervals (a, a + k) where k may take the values ±1, ±0·1, ±0·01, ±0·001, and ±0·0001.

In this situation, with the aid of the computer, we can investigate the effect of decreasing the interval in which we are asked to find the gradient of the particular chord.

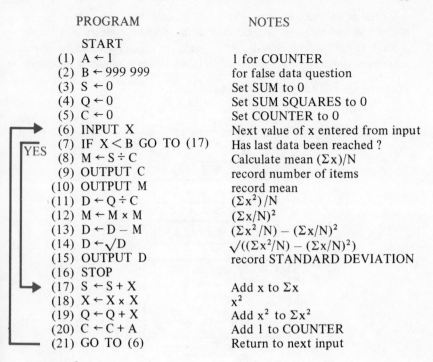

PROGRAM	NOTES
START	
(1) A ← 1	1 for COUNTER
(2) B ← 999 999	for false data question
(3) S ← 0	Set SUM to 0
(4) Q ← 0	Set SUM SQUARES to 0
(5) C ← 0	Set COUNTER to 0
(6) INPUT X	Next value of x entered from input
(7) IF X < B GO TO (17)	Has last data been reached ?
(8) M ← S ÷ C	Calculate mean $(\Sigma x)/N$
(9) OUTPUT C	record number of items
(10) OUTPUT M	record mean
(11) D ← Q ÷ C	$(\Sigma x^2)/N$
(12) M ← M × M	$(\Sigma x/N)^2$
(13) D ← D − M	$(\Sigma x^2/N) - (\Sigma x/N)^2$
(14) D ← √D	$\sqrt{((\Sigma x^2/N) - (\Sigma x/N)^2)}$
(15) OUTPUT D	record STANDARD DEVIATION
(16) STOP	
(17) S ← S + X	Add x to Σx
(18) X ← X × X	x^2
(19) Q ← Q + X	Add x^2 to Σx^2
(20) C ← C + A	Add 1 to COUNTER
(21) GO TO (6)	Return to next input

YES

program 6

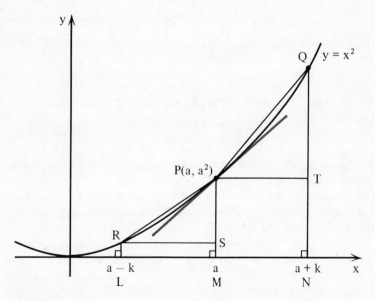

figure 10

As the interval decreases the chord will become nearer and nearer to the tangent. By considering the gradients of the chords on either side of the point P (a, a^2) we should be able to make a reasonable estimate of the gradient of the tangent at that point. In figure 10 the gradient of the chord PQ in the interval $(a, a + k)$ is given by $\dfrac{QT}{PT}$.

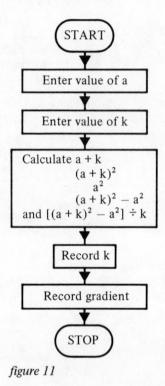

figure 11

Since QT = QN − PM and PT = MN, the gradient of the chord PQ is

$$\frac{(a + k)^2 - a^2}{k}$$

If we wished to calculate this gradient for a single value of k the flow chart would be as shown in figure 11.

As the problem requires the gradient for different values of k we can change each value after the print outs, and incorporate a conditional jump instruction to repeat the calculation.

A question like 'Is $0.00001 < k$?' would determine the correct exit from the loop. This is shown in figure 12 with each new value of k being $\frac{1}{10}$ of the previous value. The other difficulty is to carry out the calculation for the negative values of k. Here, having done so for a positive k, we could multiply k by −1 and repeat the calculation for this value.

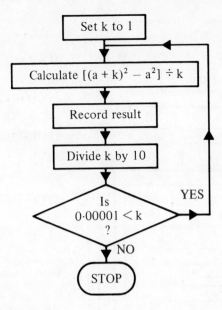

figure 12

The question 'Is $k < 0$?' would have the answer YES on the first occasion, thereby allowing the calculation to be repeated. On the second run through the answer would be NO. This is shown in figure 13. The

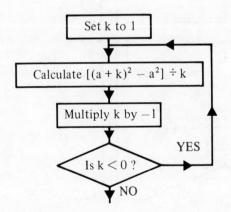

figure 13

three sections of the flow chart just covered, can now be put together into a full flow chart describing the whole problem (see figure 14 and program 7).

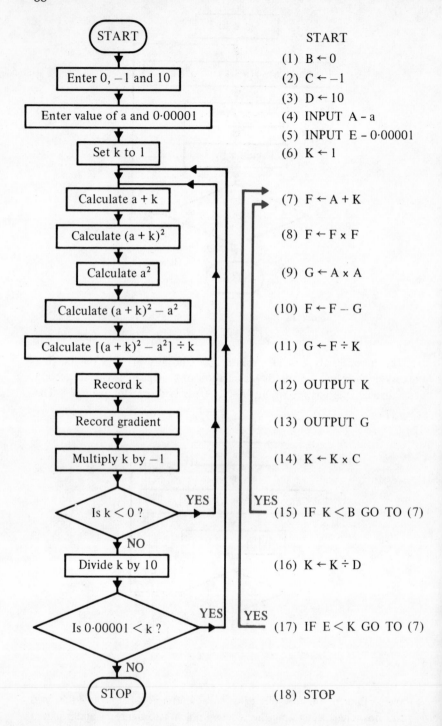

START

Enter 0, −1 and 10

Enter value of a and 0·00001

Set k to 1

Calculate a + k

Calculate (a + k)²

Calculate a²

Calculate (a + k)² − a²

Calculate [(a + k)² − a²] ÷ k

Record k

Record gradient

Multiply k by −1

Is k < 0 ? YES

NO

Divide k by 10

Is 0·00001 < k ? YES

NO

STOP

START

(1) B ← 0
(2) C ← −1
(3) D ← 10
(4) INPUT A – a
(5) INPUT E – 0·00001
(6) K ← 1

(7) F ← A + K

(8) F ← F × F

(9) G ← A × A

(10) F ← F − G

(11) G ← F ÷ K

(12) OUTPUT K

(13) OUTPUT G

(14) K ← K × C

YES
(15) IF K < B GO TO (7)

(16) K ← K ÷ D

YES
(17) IF E < K GO TO (7)

(18) STOP

figure 14   program 7

Program 7 can be adapted to calculate the gradient of the chords for any graph. In the general case, if the graph is y = f(x) the calculation steps would be finding

$$\frac{f(a + k) - f(a)}{k}$$

This is the sort of situation where a SUBROUTINE, calculating the gradient for a particular k, would be helpful. If instructions (7) to (11) were replaced by an instruction 'GO TO GRADIENT SUBROUTINE' then the main program would apply for any function.

Also we can decrease the width of the intervals merely by changing the 0·00001 in location E to some smaller degree of accuracy.

worked problem 7   Find correct to five decimal places the root of the equation $x^2 + x - 1 = 0$, which lies between 0 and 1.

There are many ways of finding the roots of mathematical equations, but perhaps the first we might meet is one based on a trial and error method.

In this situation we are told that one of the roots lies between 0 and 1. The value of $x^2 + x - 1$ is −1 when x = 0, but +1 when x = 1. This suggests that somewhere in between, the value of the expression should be 0.

We can find a better approximation for the root of the equation, by evaluating $x^2 + x - 1$ for x = 0·1, 0·2, 0·3, . . ., etc. and noting in which interval the sign of the expression changes from negative to positive. In this case $x^2 + x - 1$ is negative when x = 0·6 but positive when x = 0·7. This can then be repeated, starting at x = 0·6 and increasing in steps of 0·01 until the sign changes once again.

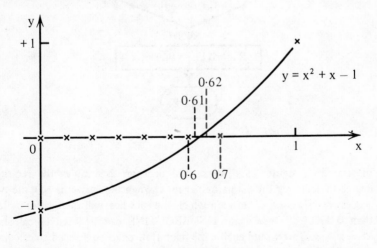

figure 15

By asking a question such as 'Is $x^2 + x - 1 < 0$?' we can repeat the calculation for the next value of x, if the answer is YES. When the answer is NO, we want to go back to the previous value of x, and increase in smaller steps. The outline flow chart in figure 16 describes the method

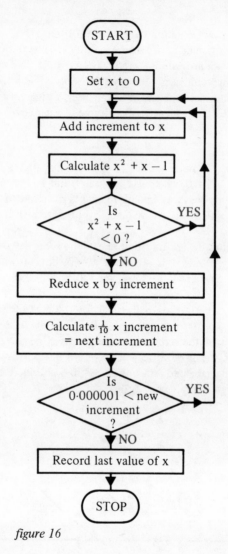

figure 16

illustrated in figure 15. The second decision determines the required degree of accuracy for the particular root. When the answer is NO, the two successive values of x, between which the root lies, will then differ by less than 0·000001. Once again a SUBROUTINE can be used to calculate $x^2 + x - 1$. This would enable the main flow chart to be used to calculate the root of any mathematical equation.

Program 8 uses a subroutine for the particular expression. The first value of x is one which makes $x^2 + x - 1$ negative, and which has been found near to the required root.

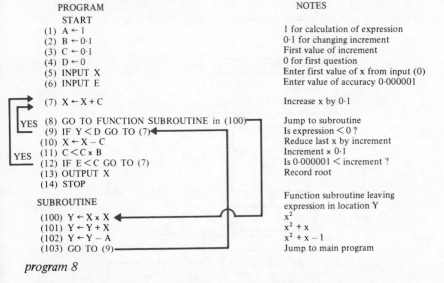

PROGRAM	NOTES
START	
(1) A ← 1	1 for calculation of expression
(2) B ← 0·1	0·1 for changing increment
(3) C ← 0·1	First value of increment
(4) D ← 0	0 for first question
(5) INPUT X	Enter first value of x from input (0)
(6) INPUT E	Enter value of accuracy 0·000001
(7) X ← X + C	Increase x by 0·1
YES (8) GO TO FUNCTION SUBROUTINE in (100)	Jump to subroutine
(9) IF Y < D GO TO (7)	Is expression < 0 ?
(10) X ← X − C	Reduce last x by increment
YES (11) C < C × B	Increment x 0·1
(12) IF E < C GO TO (7)	Is 0·000001 < increment ?
(13) OUTPUT X	Record root
(14) STOP	
SUBROUTINE	Function subroutine leaving expression in location Y
(100) Y ← X × X	x^2
(101) Y ← Y + X	$x^2 + x$
(102) Y ← Y − A	$x^2 + x - 1$
(103) GO TO (9)	Jump to main program

program 8

worked problem 8 Use the Trapezium Rule to find the area under the curve $y = x^3$ between $x = 0$ and $x = 1$. Take each interval width to be 0·1.

The area under any curve can be found approximately, by finding the area of the set of trapezia made by the chords joining two successive ordinates, see figure 17. The area of a typical trapezium PQMN is given by the formula

$$\tfrac{1}{2}(y_6 + y_7)h$$

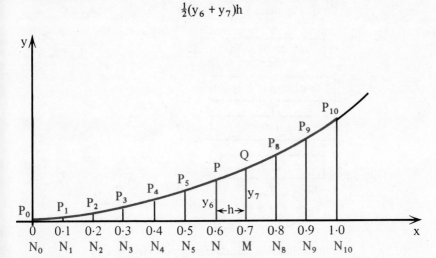

figure 17

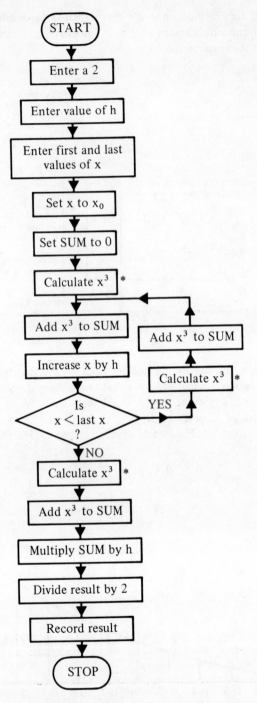

* Note that this calculation appears in three places

figure 18

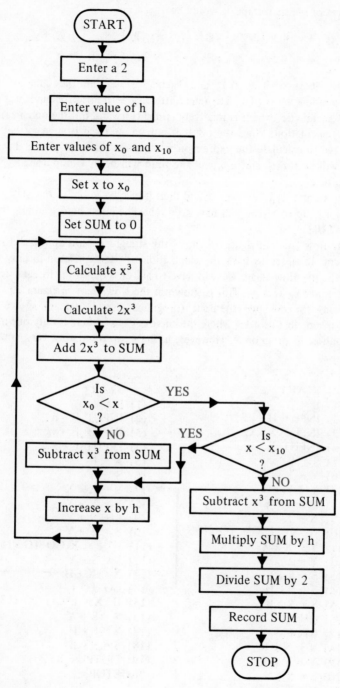

figure 19

The area of ten such trapezia would be

$$\tfrac{1}{2}(y_0 + y_1)h + \tfrac{1}{2}(y_1 + y_2)h + \tfrac{1}{2}(y_2 + y_3)h + \ldots \quad + \tfrac{1}{2}(y_9 + y_{10})h$$

$$\text{or} \quad \tfrac{1}{2}[y_0 + 2y_1 + 2y_2 + 2y_3 + \ldots + 2y_9 + y_{10}]h$$

In our situation h is 0·1, and each y would be the cube of the corresponding value of x. The main difficulty with this problem is not the actual arithmetic, which is relatively straightforward, but the organization of the calculation. Since the result is not symmetrical it appears that we will have to calculate the expression x^3, in three different places. Two of these will be for x_0 and x_{10} and the third will be in a loop using x_2, x_3, ... up to x_9.

For x_0 and x_{10}, we want to add their respective cubes to the SUM, but for x_2 ... up to x_9 we will have to double the cubes before adding them to the SUM.

The flow chart in figure 18 shows the straightforward approach to the problem. In order to have the calculation of the expression in only one place in the flow chart we will have to ask two questions to cater for x being either x_0 or x_{10}. This is shown in the flow chart in figure 19. The programs for the two methods (programs 9 and 10) are shown for comparison. In this case, where the curve is $y = x^3$, there is only one more instruction in program 9. However, if the expression were very compli-

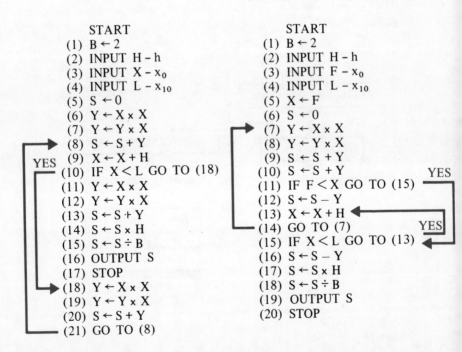

START	START
(1) B ← 2	(1) B ← 2
(2) INPUT H – h	(2) INPUT H – h
(3) INPUT X – x_0	(3) INPUT F – x_0
(4) INPUT L – x_{10}	(4) INPUT L – x_{10}
(5) S ← 0	(5) X ← F
(6) Y ← X × X	(6) S ← 0
(7) Y ← Y × X	(7) Y ← X × X
(8) S ← S + Y	(8) Y ← Y × X
(9) X ← X + H	(9) S ← S + Y
(10) IF X < L GO TO (18)	(10) S ← S + Y
(11) Y ← X × X	(11) IF F < X GO TO (15)
(12) Y ← Y × X	(12) S ← S – Y
(13) S ← S + Y	(13) X ← X + H
(14) S ← S × H	(14) GO TO (7)
(15) S ← S ÷ B	(15) IF X < L GO TO (13)
(16) OUTPUT S	(16) S ← S – Y
(17) STOP	(17) S ← S × H
(18) Y ← X × X	(18) S ← S ÷ B
(19) Y ← Y × X	(19) OUTPUT S
(20) S ← S + Y	(20) STOP
(21) GO TO (8)	

program 9   *program 10*

cated, entailing several instructions in its calculation, program 9 would be considerably longer than program 10.

In running time on the computer there would not be much to choose between the two programs. In program 10 the calculation is actually carried out three times, even though it is only written once. It does not always follow that the shorter program will necessarily be shorter to run. In the commercial world the first priority is to solve the problem or to carry out the particular job. Only then can one really afford to spend time on making the program pleasing to the programmer.

exercise 12

1 Carry out a dry check on each of programs 11 to 13 and find out what problem each is solving.

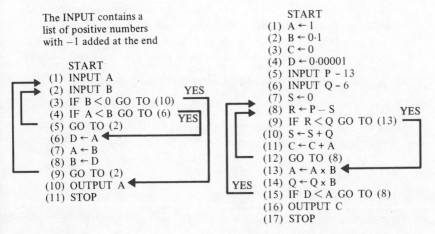

program 11   *program 12*

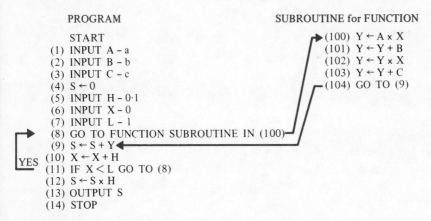

program 13

2 Grades are given for the marks obtained in an exam as follows: below 40 Grade F, 40-49 Grade E, 50-59 Grade D, 60-69 Grade C, 70-79 Grade B, 80 and over Grade A. Write a program to print out the grade for any given mark.

3 Worked problem 5, on page 83, gave one method of calculating the mean and standard deviation for a given list of numbers. Write a program to calculate the mean and standard deviation for a list of thirty numbers using the alternative formula for the S.D.

$$\sqrt{\left(\frac{N\Sigma x^2 - (\Sigma x)^2}{N(N-1)}\right)}$$

4 Write a program to arrange three given numbers in descending order. You may only use the conditional jump instruction

'IF $P < Q$ GO TO $\cdots$'

5 Worked problem 4 on page 80, printed out the investment after each year, for a regular monthly payment of £8. Write a program to find out how long, in years and months, it would take a man, investing £10 per month at 5% per annum in a similar scheme, to have saved £2000.

6 A value of π can be obtained by summing the infinite series

$$4(1 - \tfrac{1}{3} + \tfrac{1}{5} - \tfrac{1}{7} + \tfrac{1}{9} - \tfrac{1}{11} + \tfrac{1}{13} \cdots)$$

Write a program to print out the sum of the first 100 terms. Modify your program to print out the sum of 100, 200, 300, ..., 1000 terms.

appendix

running your programs on a computer

It is hoped that most of the readers of this book will have the opportunity to run their programs on a computer.

Most computers will have facilities for using one or more of the commercial languages. It would be impossible to deal with all of them, so the examples which follow are merely an indication to show the relation between these languages and the one used in the text.

The 3-address language in the book is almost directly transferable to most of the high-level languages. Whilst those high-level languages offer a wide variety of features which are extremely powerful, it is strongly recommended that in the early stages only a subset of their instructions is used. By using them in this way not only will the basic principles be grasped more quickly, but there will not be a large amount of otherwise superfluous material to confuse the issue.

As and when the 3-address form of a language is fully understood, it will then be possible to extend it to include other facilities such as compound arithmetic instructions and the more sophisticated looping procedures. Too often when introducing programming through a high-level language, the main ideas have been obscured by the punctuation that is necessary in order to get the program to run at all.

The problem that follows has been chosen because it incorporates most of the basic instructions used in the text.

problem Write a program to print out the first ten triangular numbers, using the expression $n(n-1)/2$ with $n = 1, 2, 3, \ldots, 10$.

The flow chart describing the problem is shown in figure 1 and program 1 is the corresponding 3-address program as used in the text. The examples that follow correspond as closely as possible to program 1. Some notes are included but it is assumed that the reader who wishes to use a particular language, will have the appropriate reference manual for the machine he is using.

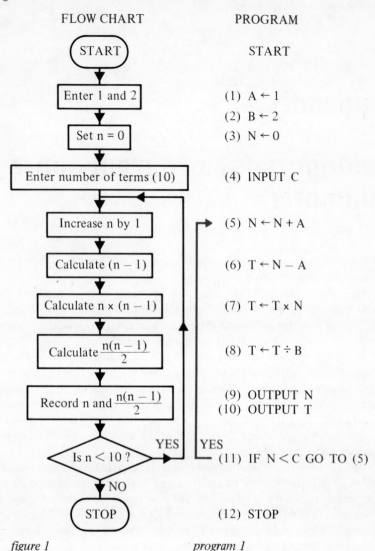

FLOW CHART	PROGRAM

START — START

Enter 1 and 2 — (1) A ← 1

(2) B ← 2

Set n = 0 — (3) N ← 0

Enter number of terms (10) — (4) INPUT C

Increase n by 1 — (5) N ← N + A

Calculate (n − 1) — (6) T ← N − A

Calculate n × (n − 1) — (7) T ← T × N

Calculate $\frac{n(n-1)}{2}$ — (8) T ← T ÷ B

Record n and $\frac{n(n-1)}{2}$ — (9) OUTPUT N
(10) OUTPUT T

Is n < 10 ? YES — (11) IF N < C GO TO (5)

NO

STOP — (12) STOP

figure 1   *program 1*

A short book list is included at the end of the appendix, but it should be pointed out that even in languages like FORTRAN and ALGOL there are many 'dialects' depending on the local installation.

Basic

In Basic each instruction is numbered 10, 20, 30, . . . etc. (see programs 2 and 3). The conditional jump will jump to the instruction with the appropriate line number. Program 3 shows the use of constants, such as 1 and 2, in the particular instruction, rather than having to designate

```
    10  LET A = 1          or    10  LET N = 0
    20  LET B = 2          ┌──▶  20  LET N = N + 1
    30  LET N = 0          │     30  LET T = N − 1
    40  INPUT C            │     40  LET T = T * N
┌──▶ 50  LET N = N + A      │     50  LET T = T/2
│    60  LET T = N − A   YES 60  PRINT N, T
│    70  LET T = T * N     └──  70  IF N < 10 THEN 20
│    80  LET T = T/B             80  END
│ YES 90  PRINT N, T
└──  100  IF N < C THEN 50
     110  END
```

program 2   *program 3*

individual stores to hold them. This is a very good language to use in the introductory stages of programming, as there is practically no punctuation.

Algol

In Algol every line must end with a semi-colon, and a declaration has to be made as to which numbers are to be integers, and which merely real. In addition not only must the program start and finish with a begin and end but so must each separate compound statement (see program 4).

Comment	Calculation of Triangular Numbers;	NOTES Title
begin	integer a, b, c, n ;	
	real T ;	Declaration of type of numbers in given locations.
	a:= 1 ;	:= means 'is set equal
	b:= 2 ;	to'.
	n:= 0 ;	
	c:= read ;	Input instruction.
L1:	n:= n + a ;	L1 is a line label for
	T:= n − a ;	the re-entry of a loop.
	T:= T * n ;	* multiply.
	T:= T/b ;	/ divide.
	write (n, T) ;	
	if n < c then goto L1;	Certain words must
end ;		be underlined.

program 4

Fortran IV

C CALCULATION OF TRIANGULAR NUMBERS   Title

```
      INTEGER A,B,C,N,T
      A = 1
      B = 2
      N = 0
      READ (1,10) C
10    FORMAT (I4)

20    N = N + A
      T = N − A
      T = T*N
      T = T/B
      WRITE (2,30) N,T
30    FORMAT (2I4)
      IF (N.LT.C) GO TO 20
      STOP
      END
```

READ (1,10) C	The 1 specifies the unit to be used for input.
10 FORMAT (I4)	The 10 indicates the type of format as specified in line 10.
WRITE (2,30) N,T / 30 FORMAT (2I4)	The 2 specifies the unit to be used for output. The 30 indicates the type of format as specified in line 30.
IF (N.LT.C) GO TO 20	(N.LT.C) is N < C
STOP	(N.GT.C) would be N > C

program 5

There are many versions of Fortran. They include a three-way conditional jump and allow the cases >, =, and < to be considered at the same time. The format statements cover the layout of data punched onto punched cards, and also that of the final print out.

Elliot 803 autocode

TITLE   CALCULATION OF TRIANGULAR .
               NUMBERS b1.   Title

```
   1)   A = 1
        B = 2
        N = 0
        READ C
   2)   N = N + A
        T = N − A
        T = T*N
        T = T/B
        PRINT N, 6
        PRINT T, 6

        JUMP IF N < C @ 2
        STOP
```

READ C	Input instruction.
PRINT N, 6 / PRINT T, 6	The 6 indicates how many digits may be required in the print out.
JUMP IF N < C @ 2	@ is a symbol on the typewriter representing 'to'.

program 6

Cobol

```
        MOVE 1 TO A .
        MOVE 2 TO B .
        MOVE 0 TO N .
        ACCEPT C .
REPEAT
        ADD A TO N GIVING N .
        SUBTRACT A FROM N GIVING T .
        MULTIPLY T BY N GIVING T .
        DIVIDE T BY B GIVING T .
        DISPLAY N .
        DISPLAY T .
        IF N LESS THAN C GO TO REPEAT .
        STOP RUN .
        END COBOL.
```

program 7

PL/I

```
/* calculation of triangular numbers */
TRIANGLE :   PROCEDURE OPTIONS (MAIN) ;
             DECLARE (A,B,C,N,T) FIXED ;
             A = 1 ;
             B = 2 ;
             N = 0 ;
             GET LIST (C) ;
BEGIN :      N = A + N ;
             T = N - A ;
             T = T * N ;
             T = T / B ;
             PUT LIST (N,T) ;
             IF N < C THEN GO TO BEGIN ;
             END ;
```

program 8

The small desk top programmable calculators such as the Olivetti Programma 101, the Hewlett Packard 9100B, or the Busicom 207 are becoming more common in educational circles. Whilst these machines are obviously not as sophisticated as a full scale computer they can offer a valuable introduction to the basic principles of programming. The language this type of machine uses is less like the 3-address language of the text than those previously mentioned, but it is not difficult to grasp.

The example in program 9 is that of the OLIVETTI PROGRAMMA 101. The set of instructions D↓, B+, D↕ is transferring n to the arithmetic unit, then adding a 1, and finally transferring the result (n + 1) back to location D.

NOTES

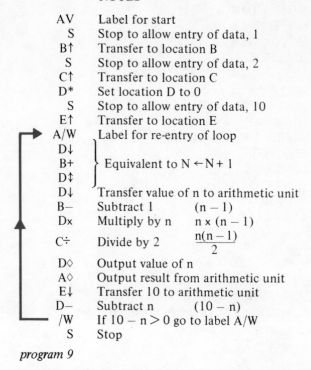

AV	Label for start
S	Stop to allow entry of data, 1
B↑	Transfer to location B
S	Stop to allow entry of data, 2
C↑	Transfer to location C
D*	Set location D to 0
S	Stop to allow entry of data, 10
E↑	Transfer to location E
A/W	Label for re-entry of loop
D↓	
B+	Equivalent to $N \leftarrow N + 1$
D↕	
D↓	Transfer value of n to arithmetic unit
B−	Subtract 1 $\quad (n - 1)$
Dx	Multiply by n $\quad n \times (n - 1)$
C÷	Divide by 2 $\quad \dfrac{n(n-1)}{2}$
D◇	Output value of n
A◇	Output result from arithmetic unit
E↓	Transfer 10 to arithmetic unit
D−	Subtract n $\quad (10 - n)$
/W	If $10 - n > 0$ go to label A/W
S	Stop

program 9

The AV is merely a coded label to indicate the start of the program whilst the /W and A/W are coded labels corresponding to the beginning and end of the loop.

book list
Basic Algol (Broderick and Barker) Iliffe.
A guide to Fortran IV programming (McCracken) Wiley.
A guide to Algol programming (McCracken) Wiley.
A guide to Cobol programming (McCracken) Wiley.
Fortran programming (Watters) Heinemann.
Students texts for Fortran IV, and PL/1 (Watters) I.B.M.
Handbooks for Fortran, Algol and Cobol (Watters) I.C.L.

solutions to exercises

exercise 1 p. 10

1(a)

A	B	C	D
12	6	~~2~~	7
~~12~~	6	13	7
42	6	13	~~7~~
42	6	~~13~~	36
42	6	6	36

(b)

A	B	C	D
19	~~8~~	21	42
~~19~~	2	21	42
2	2	~~21~~	42
2	2	4	~~42~~
2	2	4	4

2(a) 149   (b) 51   (c) 16   (d) 600

3(i) a, c, d, e, f, h   (ii) a, c, e

exercise 2 p. 12

1(a) 2   (b) 350   (c) 1, 2   (d) 14

2(a)
C ← C + D
A ← A × B
A ← A + C
OUTPUT A

(b)
A ← A + B
A ← A + C
A ← A + D
OUTPUT A

(c)
C ← C + B
C ← A + C
C ← C ÷ D
OUTPUT C

(d)
A ← A × A
D ← D + D
A ← A × D
OUTPUT A

3(a)
E ← A + B
E ← E × D
OUTPUT E

(b)
E ← B × D
E ← A + E
OUTPUT E

(c)
B ← B + D
A ← A + C
A ← B ÷ A
OUTPUT A

project p. 13

1   37, 13

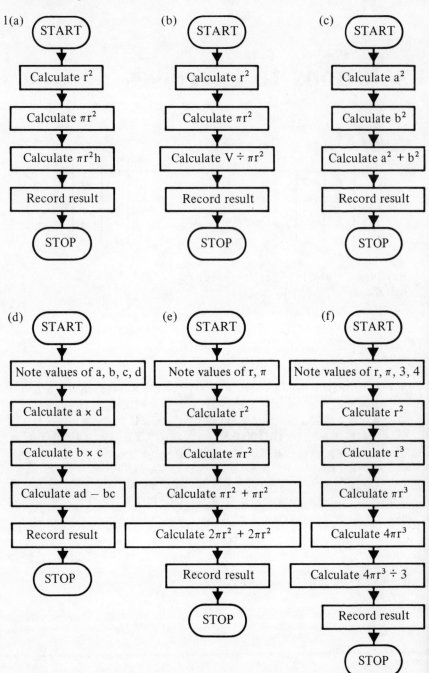

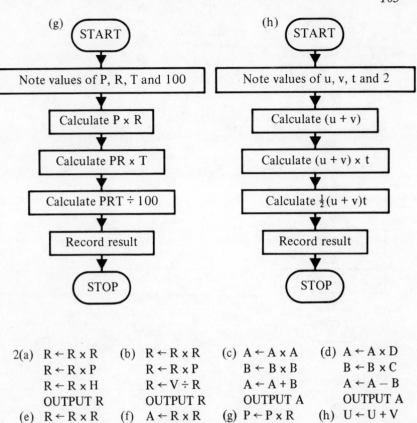

2(a) R ← R × R   (b) R ← R × R   (c) A ← A × A   (d) A ← A × D
     R ← R × P   R ← R × P   B ← B × B   B ← B × C
     R ← R × H   R ← V ÷ R   A ← A + B   A ← A − B
     OUTPUT R   OUTPUT R   OUTPUT A   OUTPUT A

(e) R ← R × R   (f) A ← R × R   (g) P ← P × R   (h) U ← U + V
     R ← R × P   A ← R × A   P ← P × T   U ← U × T
     R ← R + R   A ← A × P   P ← P ÷ H   U ← U ÷ B
     R ← R + R   A ← A × F   OUTPUT P   OUTPUT U
     OUTPUT R   A ← A ÷ T
                       OUTPUT A

3

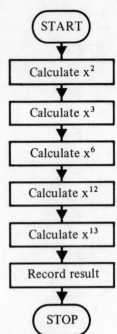

PROGRAM
$Y \leftarrow X \times X$
$Y \leftarrow Y \times X$
$Y \leftarrow Y \times Y$
$Y \leftarrow Y \times Y$
$Y \leftarrow Y \times X$
OUTPUT Y

4   6 instructions
     2 stores

project p. 19
1·5

exercise 4 p. 22

1(a) START	(b) START	(c) START
INPUT 2 to D	INPUT 1 to A	INPUT a to A
INPUT a to A	INPUT u to U	$C \leftarrow A \times A$
INPUT b to B	INPUT v to V	$C \leftarrow C \times A$
INPUT h to H	$B \leftarrow A \div U$	INPUT b to B
$C \leftarrow A + B$	$C \leftarrow A \div V$	$D \leftarrow B \times B$
$H \leftarrow H + H$	$B \leftarrow B + C$	$D \leftarrow D \times B$
$C \leftarrow C \div H$	$B \leftarrow A \div B$	$C \leftarrow C + D$
OUTPUT C	OUTPUT B	OUTPUT C
STOP	STOP	STOP

2   For program 4:   For program 5:   For program 6:

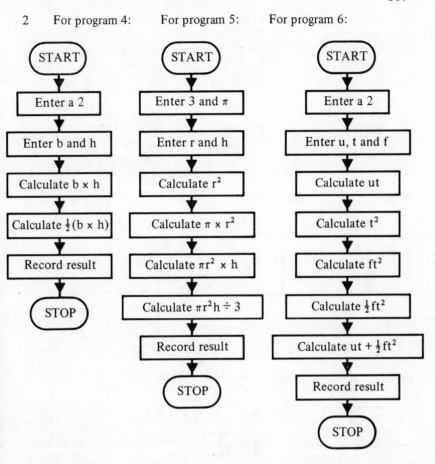

project p. 25

1   36

exercise 5 p. 31

1 program 5: $\dfrac{x - y}{xy}$ program 6: $x + y + z$ program 7: $xy + z$

program 8: $\dfrac{xz + yz}{xy}$ program 9: $\dfrac{x^2 + xy + y^2}{x + y}$ program 10: $\dfrac{xz + yz}{3xy}$

All except program 6.

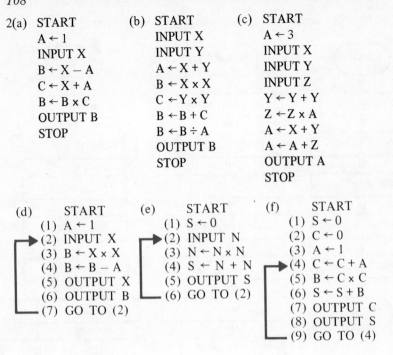

2(a) START
A ← 1
INPUT X
B ← X − A
C ← X + A
B ← B × C
OUTPUT B
STOP

(b) START
INPUT X
INPUT Y
A ← X + Y
B ← X × X
C ← Y × Y
B ← B + C
B ← B ÷ A
OUTPUT B
STOP

(c) START
A ← 3
INPUT X
INPUT Y
INPUT Z
Y ← Y + Y
Z ← Z × A
A ← X + Y
A ← A + Z
OUTPUT A
STOP

(d) START
(1) A ← 1
(2) INPUT X
(3) B ← X × X
(4) B ← B − A
(5) OUTPUT X
(6) OUTPUT B
(7) GO TO (2)

(e) START
(1) S ← 0
(2) INPUT N
(3) N ← N × N
(4) S ← N + N
(5) OUTPUT S
(6) GO TO (2)

(f) START
(1) S ← 0
(2) C ← 0
(3) A ← 1
(4) C ← C + A
(5) B ← C × C
(6) S ← S + B
(7) OUTPUT C
(8) OUTPUT S
(9) GO TO (4)

exercise 6 p. 34

1

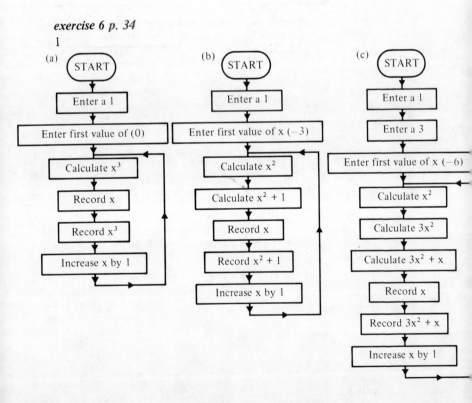

(a) START
Enter a 1
Enter first value of (0)
Calculate x^3
Record x
Record x^3
Increase x by 1

(b) START
Enter a 1
Enter first value of x (−3)
Calculate x^2
Calculate $x^2 + 1$
Record x
Record $x^2 + 1$
Increase x by 1

(c) START
Enter a 1
Enter a 3
Enter first value of x (−6)
Calculate x^2
Calculate $3x^2$
Calculate $3x^2 + x$
Record x
Record $3x^2 + x$
Increase x by 1

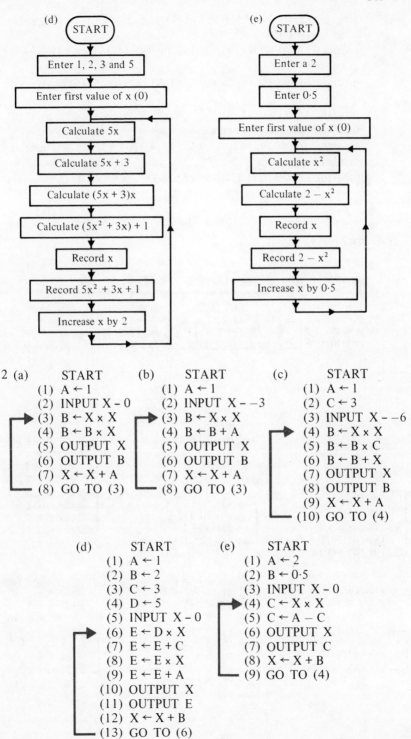

(d) START

Enter 1, 2, 3 and 5

Enter first value of x (0)

Calculate 5x

Calculate 5x + 3

Calculate (5x + 3)x

Calculate $(5x^2 + 3x) + 1$

Record x

Record $5x^2 + 3x + 1$

Increase x by 2

(e) START

Enter a 2

Enter 0·5

Enter first value of x (0)

Calculate x^2

Calculate $2 - x^2$

Record x

Record $2 - x^2$

Increase x by 0·5

2 (a)   START
    (1) A ← 1
    (2) INPUT X − 0
    (3) B ← X × X
    (4) B ← B × X
    (5) OUTPUT X
    (6) OUTPUT B
    (7) X ← X + A
    (8) GO TO (3)

(b)   START
    (1) A ← 1
    (2) INPUT X − −3
    (3) B ← X × X
    (4) B ← B + A
    (5) OUTPUT X
    (6) OUTPUT B
    (7) X ← X + A
    (8) GO TO (3)

(c)   START
    (1) A ← 1
    (2) C ← 3
    (3) INPUT X − −6
    (4) B ← X × X
    (5) B ← B × C
    (6) B ← B + X
    (7) OUTPUT X
    (8) OUTPUT B
    (9) X ← X + A
    (10) GO TO (4)

(d)   START
    (1) A ← 1
    (2) B ← 2
    (3) C ← 3
    (4) D ← 5
    (5) INPUT X − 0
    (6) E ← D × X
    (7) E ← E + C
    (8) E ← E × X
    (9) E ← E + A
    (10) OUTPUT X
    (11) OUTPUT E
    (12) X ← X + B
    (13) GO TO (6)

(e)   START
    (1) A ← 2
    (2) B ← 0·5
    (3) INPUT X − 0
    (4) C ← X × X
    (5) C ← A − C
    (6) OUTPUT X
    (7) OUTPUT C
    (8) X ← X + B
    (9) GO TO (4)

3   program 12: (0, 0), (0·5, 0·25), (1, 1), (1·5, 2·25), (2, 4), (2·5, 6·25)
     − x^2 in steps of $\frac{1}{2}$.

program 13: (0, 7), (1, 9), (2, 11), (3, 13), (4, 15), (5, 17) − 2x + 7
in steps of 1

program 14: (0, 0), (1, 1), (2, 3), (3, 6), (4, 10), (5, 15) − $\frac{1}{2}$x(x + 1)
in steps of 1

program 15: (1, 1, 1), (1, 2, 0·5), (2, 3, 0·66), (3, 5, 0·6), (5, 8,
0·625) − Fibonacci numbers and ratios

program 16: (1, 1), (2, 3), (3, 6), (4, 10), (5, 15) − Triangular
numbers

program 17: 0·5, 0·667, 0·75, 0·8, 0·833 − x ÷ (x + 1) in steps of 1

exercise 7 p. 38

1

(a)   START
   (1) A ← 1
   (2) T ← 1
→ (3) R ← T × T
→ (4) R ← R × T
   (5) OUTPUT R
   (6) T ← T + A
└─ (7) GO TO (3)

(b)   START
   (1) B ← 2
   (2) T ← 1
→ (3) T ← T × B
   (4) OUTPUT T
└─ (5) GO TO (3)

(c)   START
   (1) A ← 1
   (2) B ← 2
→ (3) T ← A ÷ B
   (4) OUTPUT T
└─ (5) GO TO (3)

(d)   START
   (1) A ← 1
   (2) T ← 1
   (3) N ← 1
→ (4) OUTPUT T
   (5) N ← N + A
   (6) T ← T ÷ N
└─ (7) GO TO (4)

2

(a)   START
   (1) A ← 1
   (2) N ← 1
   (3) S ← 1
→ (4) N ← N + A
   (5) T ← N × N
   (6) T ← T × N
   (7) S ← S + T
   (8) OUTPUT S
└─ (9) GO TO (4)

(b)   START
   (1) A ← 1
   (2) B ← 3
   (3) S ← 1
   (4) T ← A ÷ B
→ (5) S ← S + T
   (6) OUTPUT S
   (7) T ← T ÷ B
└─ (8) GO TO (5)

(c)   START
   (1) A ← 2
   (2) S ← 2
   (3) T ← S × A
→ (4) S ← S + T
   (5) OUTPUT S
   (6) T ← T × A
└─ (7) GO TO (4)

(d)   START
   (1) A ← 2
   (2) T ← 1
   (3) S ← 1
→ (4) T ← T + A
   (5) S ← S + T
   (6) OUTPUT S
└─ (7) GO TO (4)

(e)   START
   (1) A ← 1
   (2) T ← 1
   (3) N ← 1
   (4) S ← 1
→ (5) T ← T ÷ N
   (6) S ← S + T
   (7) OUTPUT S
   (8) N ← N + A
└─ (9) GO TO (5)

(f)   START
   (1) A ← 2
   (2) S ← 1
   (3) T ← S ÷ A
→ (4) S ← S + T
   (5) OUTPUT S
   (6) T ← T ÷ A
└─ (7) GO TO 4

exercise 8 p. 47

1(a)

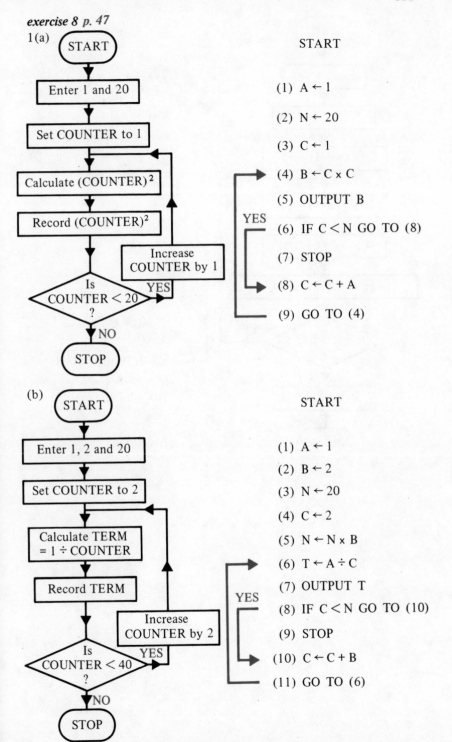

START

Enter 1 and 20

Set COUNTER to 1

Calculate (COUNTER)²

Record (COUNTER)²

Increase COUNTER by 1

Is COUNTER < 20 ? — YES

NO

STOP

START

(1) A ← 1

(2) N ← 20

(3) C ← 1

(4) B ← C × C

(5) OUTPUT B

YES

(6) IF C < N GO TO (8)

(7) STOP

(8) C ← C + A

(9) GO TO (4)

(b)

START

Enter 1, 2 and 20

Set COUNTER to 2

Calculate TERM = 1 ÷ COUNTER

Record TERM

Increase COUNTER by 2

Is COUNTER < 40 ? — YES

NO

STOP

START

(1) A ← 1

(2) B ← 2

(3) N ← 20

(4) C ← 2

(5) N ← N × B

(6) T ← A ÷ C

(7) OUTPUT T

YES

(8) IF C < N GO TO (10)

(9) STOP

(10) C ← C + B

(11) GO TO (6)

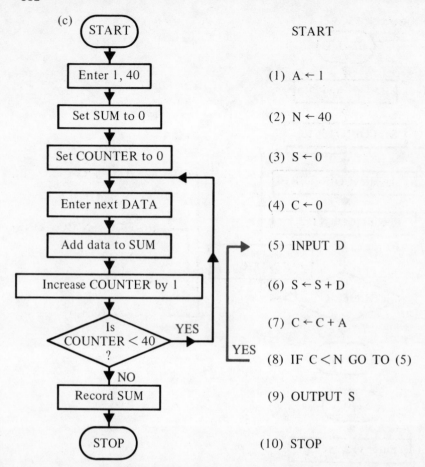

(c)

START

Enter 1, 40

Set SUM to 0

Set COUNTER to 0

Enter next DATA

Add data to SUM

Increase COUNTER by 1

Is COUNTER < 40 ?

YES

NO

Record SUM

STOP

START

(1) $A \leftarrow 1$

(2) $N \leftarrow 40$

(3) $S \leftarrow 0$

(4) $C \leftarrow 0$

(5) INPUT D

(6) $S \leftarrow S + D$

(7) $C \leftarrow C + A$

(8) IF $C < N$ GO TO (5)

(9) OUTPUT S

(10) STOP

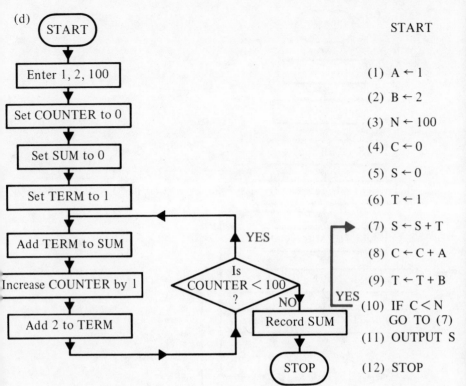

(d)

START

Enter 1, 2, 100

Set COUNTER to 0

Set SUM to 0

Set TERM to 1

Add TERM to SUM

Increase COUNTER by 1

Add 2 to TERM

Is COUNTER < 100 ?

YES

NO

Record SUM

STOP

START

(1) A ← 1

(2) B ← 2

(3) N ← 100

(4) C ← 0

(5) S ← 0

(6) T ← 1

(7) S ← S + T

(8) C ← C + A

(9) T ← T + B

(10) IF C < N GO TO (7)

(11) OUTPUT S

(12) STOP

YES

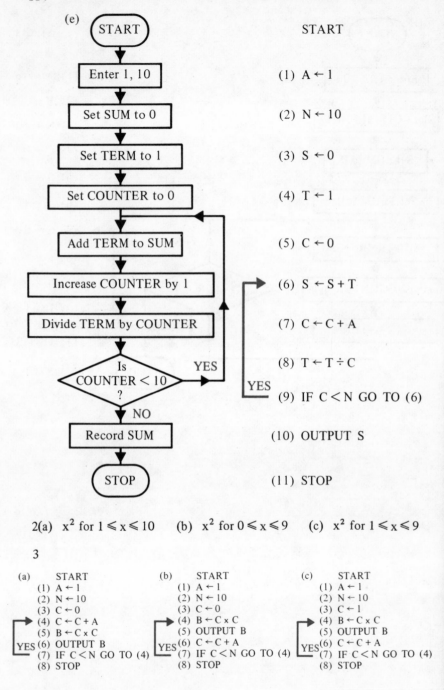

(e)

START

Enter 1, 10

Set SUM to 0

Set TERM to 1

Set COUNTER to 0

Add TERM to SUM

Increase COUNTER by 1

Divide TERM by COUNTER

Is COUNTER < 10 ? YES

NO

Record SUM

STOP

START

(1) $A \leftarrow 1$

(2) $N \leftarrow 10$

(3) $S \leftarrow 0$

(4) $T \leftarrow 1$

(5) $C \leftarrow 0$

(6) $S \leftarrow S + T$

(7) $C \leftarrow C + A$

(8) $T \leftarrow T \div C$

(9) IF $C < N$ GO TO (6)

(10) OUTPUT S

(11) STOP

2(a) x^2 for $1 \leqslant x \leqslant 10$ (b) x^2 for $0 \leqslant x \leqslant 9$ (c) x^2 for $1 \leqslant x \leqslant 9$

3

(a) START
(1) $A \leftarrow 1$
(2) $N \leftarrow 10$
(3) $C \leftarrow 0$
(4) $C \leftarrow C + A$
(5) $B \leftarrow C \times C$
YES (6) OUTPUT B
(7) IF $C < N$ GO TO (4)
(8) STOP

(b) START
(1) $A \leftarrow 1$
(2) $N \leftarrow 10$
(3) $C \leftarrow 0$
(4) $B \leftarrow C \times C$
(5) OUTPUT B
YES (6) $C \leftarrow C + A$
(7) IF $C < N$ GO TO (4)
(8) STOP

(c) START
(1) $A \leftarrow 1$
(2) $N \leftarrow 10$
(3) $C \leftarrow 1$
(4) $B \leftarrow C \times C$
(5) OUTPUT B
YES (6) $C \leftarrow C + A$
(7) IF $C < N$ GO TO (4)
(8) STOP

exercise 9 p. 54

1(a) x^2 for $1 \leqslant x \leqslant 11$ (b) x^2 for $0 \leqslant x \leqslant 9$

2

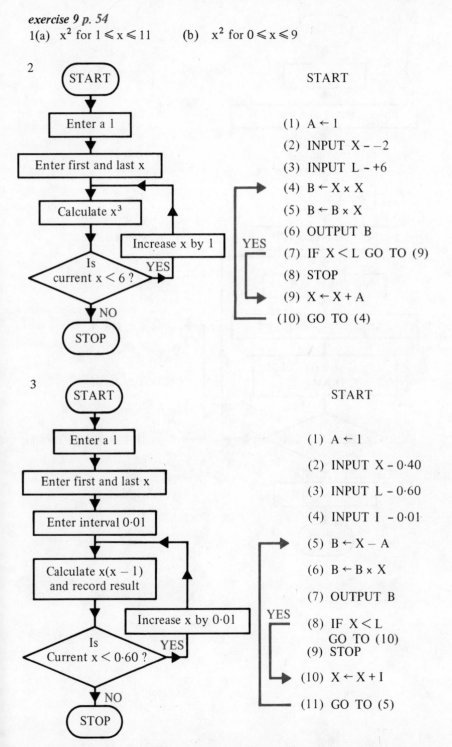

START

(1) $A \leftarrow 1$

(2) INPUT $X - -2$

(3) INPUT $L - +6$

(4) $B \leftarrow X \times X$

(5) $B \leftarrow B \times X$

(6) OUTPUT B

(7) IF $X < L$ GO TO (9)

(8) STOP

(9) $X \leftarrow X + A$

(10) GO TO (4)

3

START

(1) $A \leftarrow 1$

(2) INPUT $X - 0 \cdot 40$

(3) INPUT $L - 0 \cdot 60$

(4) INPUT $I - 0 \cdot 01$

(5) $B \leftarrow X - A$

(6) $B \leftarrow B \times X$

(7) OUTPUT B

(8) IF $X < L$ GO TO (10)

(9) STOP

(10) $X \leftarrow X + I$

(11) GO TO (5)

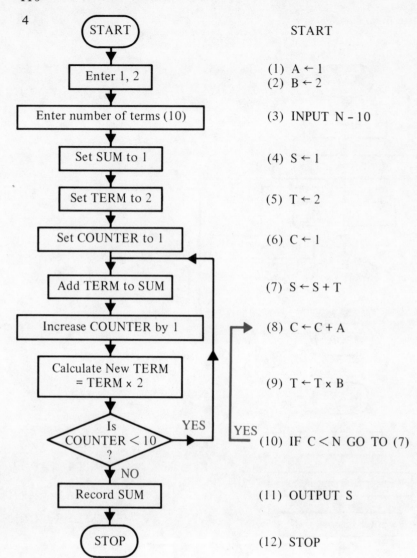

START

(1) A ← 1
(2) B ← 2

(3) INPUT N - 10

(4) S ← 1

(5) T ← 2

(6) C ← 1

(7) S ← S + T

(8) C ← C + A

(9) T ← T × B

(10) IF C < N GO TO (7)

(11) OUTPUT S

(12) STOP

5

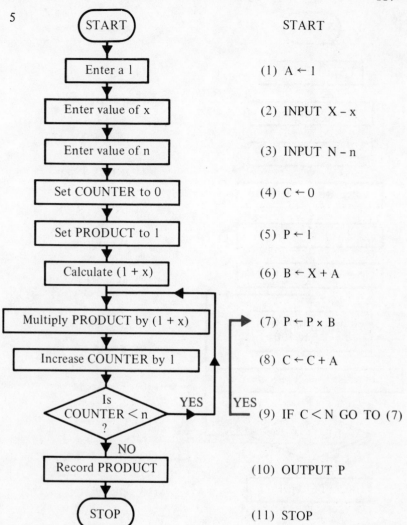

START

(1) A ← 1

(2) INPUT X – x

(3) INPUT N – n

(4) C ← 0

(5) P ← 1

(6) B ← X + A

(7) P ← P × B

(8) C ← C + A

(9) IF C < N GO TO (7)

(10) OUTPUT P

(11) STOP

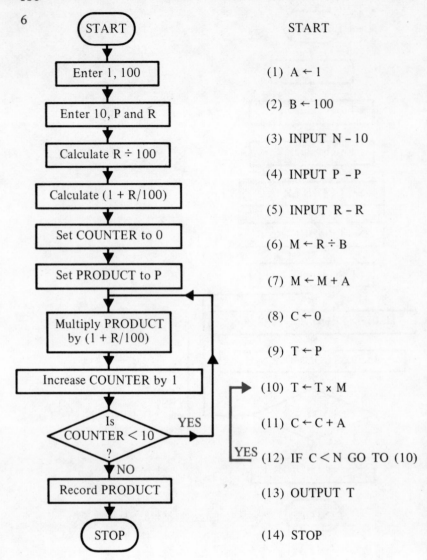

START

(1) A ← 1

(2) B ← 100

(3) INPUT N – 10

(4) INPUT P – P

(5) INPUT R – R

(6) M ← R ÷ B

(7) M ← M + A

(8) C ← 0

(9) T ← P

(10) T ← T × M

(11) C ← C + A

YES (12) IF C < N GO TO (10)

(13) OUTPUT T

(14) STOP

7

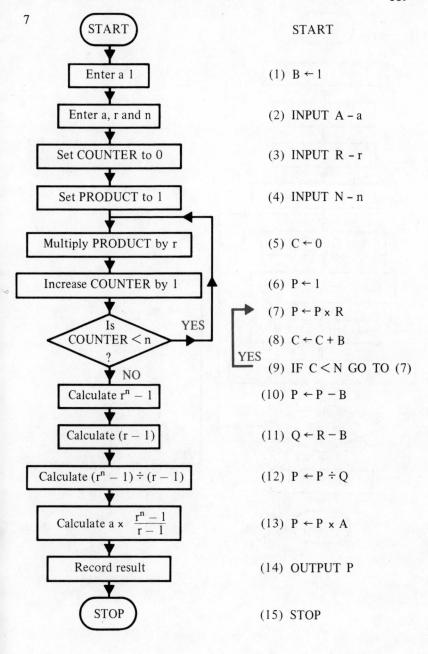

START

(1) B ← 1

(2) INPUT A – a

(3) INPUT R – r

(4) INPUT N – n

(5) C ← 0

(6) P ← 1

(7) P ← P × R

(8) C ← C + B

(9) IF C < N GO TO (7)

(10) P ← P – B

(11) Q ← R – B

(12) P ← P ÷ Q

(13) P ← P × A

(14) OUTPUT P

(15) STOP

8

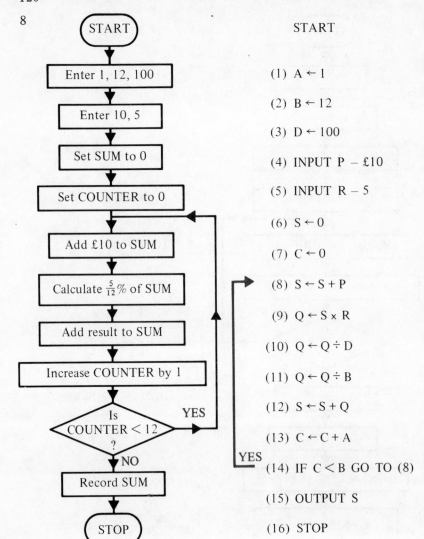

START

START

Enter 1, 12, 100

(1) $A \leftarrow 1$

Enter 10, 5

(2) $B \leftarrow 12$

Set SUM to 0

(3) $D \leftarrow 100$

Set COUNTER to 0

(4) INPUT P – £10

Add £10 to SUM

(5) INPUT R – 5

Calculate $\frac{5}{12}$% of SUM

(6) $S \leftarrow 0$

Add result to SUM

(7) $C \leftarrow 0$

Increase COUNTER by 1

(8) $S \leftarrow S + P$

Is COUNTER < 12 ?   YES

(9) $Q \leftarrow S \times R$

(10) $Q \leftarrow Q \div D$

(11) $Q \leftarrow Q \div B$

(12) $S \leftarrow S + Q$

(13) $C \leftarrow C + A$

NO

YES

Record SUM

(14) IF $C < B$ GO TO (8)

(15) OUTPUT S

STOP

(16) STOP

9

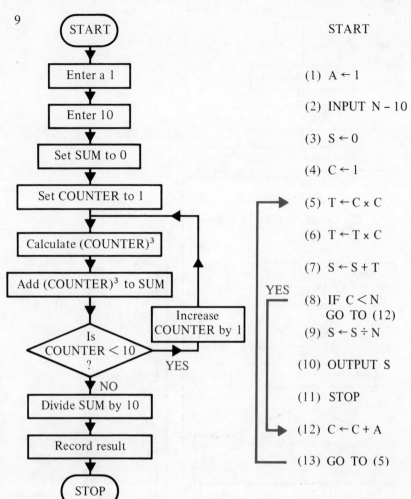

START

(1) $A \leftarrow 1$

(2) INPUT $N - 10$

(3) $S \leftarrow 0$

(4) $C \leftarrow 1$

(5) $T \leftarrow C \times C$

(6) $T \leftarrow T \times C$

(7) $S \leftarrow S + T$

(8) IF $C < N$
GO TO (12)

(9) $S \leftarrow S \div N$

(10) OUTPUT S

(11) STOP

(12) $C \leftarrow C + A$

(13) GO TO (5)

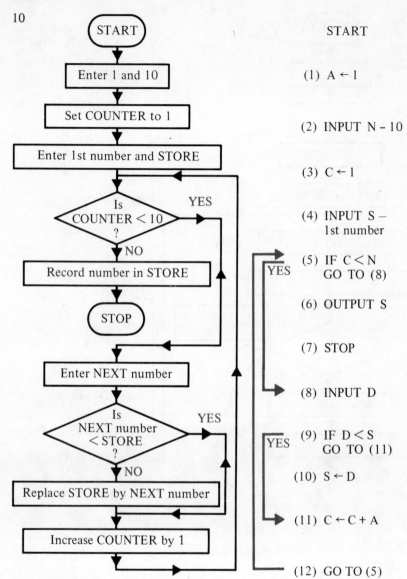

START

(1) A ← 1

(2) INPUT N – 10

(3) C ← 1

(4) INPUT S –
    1st number

(5) IF C < N
    GO TO (8)

(6) OUTPUT S

(7) STOP

(8) INPUT D

(9) IF D < S
    GO TO (11)

(10) S ← D

(11) C ← C + A

(12) GO TO (5)

project p. 60

The program is finding the average of the squares of the numbers entered from the paper tape.

project p. 67

The subroutine calculates X^5. When used with the main program OUTPUT X is 10, and OUTPUT Y is the sum of the first ten fifth powers. The second subroutine calculates $1/X^6$ and the outputs are again 10, and the sum of the reciprocals of the first ten sixth powers.

exercise 10 p. 69

1

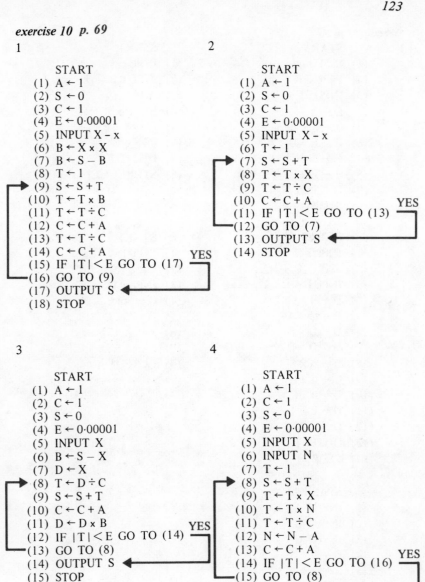

```
        START
    (1) A ← 1
    (2) S ← 0
    (3) C ← 1
    (4) E ← 0·00001
    (5) INPUT X – x
    (6) B ← X × X
    (7) B ← S – B
    (8) T ← 1
  ► (9) S ← S + T
   (10) T ← T × B
   (11) T ← T ÷ C
   (12) C ← C + A
   (13) T ← T ÷ C
   (14) C ← C + A
   (15) IF |T| < E GO TO (17)        YES
  ─(16) GO TO (9)
   (17) OUTPUT S  ◄
   (18) STOP
```

2

```
        START
    (1) A ← 1
    (2) S ← 0
    (3) C ← 1
    (4) E ← 0·00001
    (5) INPUT X – x
    (6) T ← 1
  ► (7) S ← S + T
    (8) T ← T × X
    (9) T ← T ÷ C
   (10) C ← C + A
   (11) IF |T| < E GO TO (13)        YES
  ─(12) GO TO (7)
   (13) OUTPUT S  ◄
   (14) STOP
```

3

```
        START
    (1) A ← 1
    (2) C ← 1
    (3) S ← 0
    (4) E ← 0·00001
    (5) INPUT X
    (6) B ← S – X
    (7) D ← X
  ► (8) T ← D ÷ C
    (9) S ← S + T
   (10) C ← C + A
   (11) D ← D × B
   (12) IF |T| < E GO TO (14)        YES
  ─(13) GO TO (8)
   (14) OUTPUT S  ◄
   (15) STOP
```

4

```
        START
    (1) A ← 1
    (2) C ← 1
    (3) S ← 0
    (4) E ← 0·00001
    (5) INPUT X
    (6) INPUT N
    (7) T ← 1
  ► (8) S ← S + T
    (9) T ← T × X
   (10) T ← T × N
   (11) T ← T ÷ C
   (12) N ← N – A
   (13) C ← C + A
   (14) IF |T| < E GO TO (16)        YES
  ─(15) GO TO (8)
   (16) OUTPUT S  ◄
   (17) STOP
```

project p. 70

1st **OUTPUT** $4·3 \times 2·7 \times \sin 30 \simeq 5·8$

2nd **OUTPUT** $4·3 \times 2·7 \times \sin 50 \simeq 8·9$

124

exercise 11 p. 71

1   START
    (1) INPUT A – a
    (2) INPUT B – b
    (3) INPUT C – $C°$
    (4) $D \leftarrow 2$
    (5) $E \leftarrow A \times B$
    (6) $E \leftarrow E \div D$
    (7) $F \leftarrow \text{Sin } C$
    (8) $E \leftarrow E \times F$
    (9) OUTPUT E
  (10) STOP

2   START
    (1) INPUT A – a
    (2) INPUT B – b
    (3) $A \leftarrow A \times A$
    (4) $B \leftarrow B \times B$
    (5) $C \leftarrow A + B$
    (6) $C \leftarrow \sqrt{C}$
    (7) OUTPUT C
    (8) STOP

3   START
    (1) INPUT B – b
    (2) INPUT C – c
    (3) INPUT A – $A°$
    (4) $D \leftarrow B \times C$
    (5) $E \leftarrow \text{Cos } A$
    (6) $D \leftarrow D \times E$
    (7) $D \leftarrow D \times D$
    (8) $B \leftarrow B \times B$
    (9) $C \leftarrow C \times C$
  (10) $E \leftarrow B + C$
  (11) $D \leftarrow E - D$
  (12) $D \leftarrow \sqrt{D}$
  (13) OUTPUT D
  (14) STOP

4   START
    (1) INPUT A – a
    (2) INPUT C – $A°$
    (3) INPUT D – $B°$
    (4) $C \leftarrow \text{Sin } C$
    (5) $D \leftarrow \text{Sin } D$
    (6) $A \leftarrow A \div C$
    (7) $A \leftarrow A \times D$
    (8) OUTPUT A
    (9) STOP

5   START
    (1) INPUT A
    (2) INPUT B
    (3) INPUT C
    (4) $D \leftarrow 2$
    (5) $S \leftarrow A + B$
    (6) $S \leftarrow S + C$
    (7) $S \leftarrow S \div D$
    (8) $E \leftarrow S - A$
    (9) $F \leftarrow S - B$
  (10) $E \leftarrow E \times F$
  (11) $F \leftarrow S - C$
  (12) $E \leftarrow E \times F$
  (13) $S \leftarrow S \times E$
  (14) $S \leftarrow \sqrt{S}$
  (15) OUTPUT S
  (16) STOP

exercise 12 p. 95

1 program 11: The problem is to find the largest number in a given list of positive numbers.

program 12: 13 ÷ 6 with the result correct to 4 decimal places.

program 13: The subroutine evaluates $ax^2 + bx + c$ by nested multiplication. The main program is finding an approximation to the area under the curve $y = ax^2 + bx + c$. The method used is to sum the rectangles under the curve for each of the 10 intervals from $x = 0$ to $x = 1$. The formula for this approximation to the area is

$$(y_0 + y_1 + y_2 + \ldots + y_9) \times h.$$

2

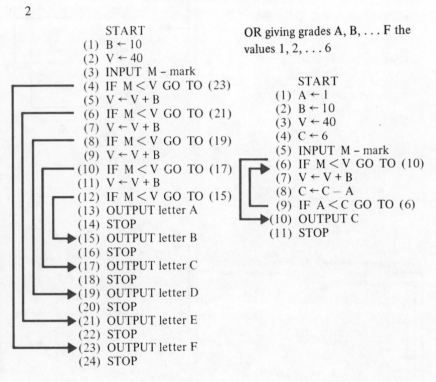

```
            START                    OR giving grades A, B, ... F the
      (1)  B ← 10                    values 1, 2, ... 6
      (2)  V ← 40
      (3)  INPUT  M – mark
      (4)  IF  M < V  GO TO (23)              START
      (5)  V ← V + B                   (1)  A ← 1
      (6)  IF  M < V  GO TO (21)       (2)  B ← 10
      (7)  V ← V + B                   (3)  V ← 40
      (8)  IF  M < V  GO TO (19)       (4)  C ← 6
      (9)  V ← V + B                   (5)  INPUT  M – mark
     (10)  IF  M < V  GO TO (17)       (6)  IF  M < V  GO TO (10)
     (11)  V ← V + B                   (7)  V ← V + B
     (12)  IF  M < V  GO TO (15)       (8)  C ← C – A
     (13)  OUTPUT letter A             (9)  IF  A < C  GO TO (6)
     (14)  STOP                       (10)  OUTPUT C
     (15)  OUTPUT letter B            (11)  STOP
     (16)  STOP
     (17)  OUTPUT letter C
     (18)  STOP
     (19)  OUTPUT letter D
     (20)  STOP
     (21)  OUTPUT letter E
     (22)  STOP
     (23)  OUTPUT letter F
     (24)  STOP
```

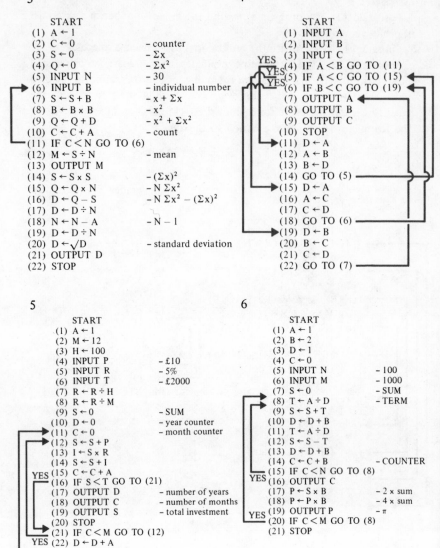

3

```
        START
(1)  A ← 1
(2)  C ← 0                    - counter
(3)  S ← 0                    - Σx
(4)  Q ← 0                    - Σx²
(5)  INPUT N                  - 30
(6)  INPUT B                  - individual number
(7)  S ← S + B               - x + Σx
(8)  B ← B × B               - x²
(9)  Q ← Q + D               - x² + Σx²
(10) C ← C + A               - count
(11) IF C < N GO TO (6)
(12) M ← S ÷ N               - mean
(13) OUTPUT M
(14) S ← S × S               - (Σx)²
(15) Q ← Q × N               - N Σx²
(16) D ← Q − S               - N Σx² − (Σx)²
(17) D ← D ÷ N
(18) N ← N − A               - N − 1
(19) D ← D ÷ N
(20) D ← √D                  - standard deviation
(21) OUTPUT D
(22) STOP
```

4

```
        START
(1)  INPUT A
(2)  INPUT B
(3)  INPUT C
(4)  IF A < B GO TO (11)      YES
(5)  IF A < C GO TO (15)      YES
(6)  IF B < C GO TO (19)      YES
(7)  OUTPUT A
(8)  OUTPUT B
(9)  OUTPUT C
(10) STOP
(11) D ← A
(12) A ← B
(13) B ← D
(14) GO TO (5)
(15) D ← A
(16) A ← C
(17) C ← D
(18) GO TO (6)
(19) D ← B
(20) B ← C
(21) C ← D
(22) GO TO (7)
```

5

```
        START
(1)  A ← 1
(2)  M ← 12
(3)  H ← 100
(4)  INPUT P                  - £10
(5)  INPUT R                  - 5%
(6)  INPUT T                  - £2000
(7)  R ← R ÷ H
(8)  R ← R ÷ M
(9)  S ← 0                    - SUM
(10) D ← 0                    - year counter
(11) C ← 0                    - month counter
(12) S ← S + P
(13) I ← S × R
(14) S ← S + I
(15) C ← C + A
(16) IF S < T GO TO (21)      YES
(17) OUTPUT D                 - number of years
(18) OUTPUT C                 - number of months
(19) OUTPUT S                 - total investment
(20) STOP
(21) IF C < M GO TO (12)
(22) D ← D + A               YES
(23) GO TO (11)
```

6

```
        START
(1)  A ← 1
(2)  B ← 2
(3)  D ← 1
(4)  C ← 0
(5)  INPUT N                  - 100
(6)  INPUT M                  - 1000
(7)  S ← 0                    - SUM
(8)  T ← A ÷ D               - TERM
(9)  S ← S + T
(10) D ← D + B
(11) T ← A ÷ D
(12) S ← S − T
(13) D ← D + B
(14) C ← C + B               - COUNTER
(15) IF C < N GO TO (8)       YES
(16) OUTPUT C
(17) P ← S × B               - 2 × sum
(18) P ← P × B               - 4 × sum
(19) OUTPUT P                 - π
(20) IF C < M GO TO (8)       YES
(21) STOP
```

index